The
Golden Ratio
and Fibonacci
Numbers

The Golden Ratio _and_ Fibonacci Numbers

by

Richard A. Dunlap

Dalhousie University
Canada

World Scientific
New Jersey • London • Singapore • Hong Kong

Published by

World Scientific Publishing Co. Pte. Ltd.

5 Toh Tuck Link, Singapore 596224

USA office: 27 Warren Street, Suite 401-402, Hackensack, NJ 07601

UK office: 57 Shelton Street, Covent Garden, London WC2H 9HE

Library of Congress Cataloging-in-Publication Data
Dunlap, R. A.
   The golden ratio and Fibonacci numbers / R. A. Dunlap.
     p.   cm.
   Includes bibliographical references (pp. 153–155) and index.
   ISBN-13 978-981-02-3264-1
   ISBN-10 981-02-3264-0 (alk. paper)
     1. Golden section.   2. Fibonacci numbers.   I. Title.
   QA466.D86   1997
   512'.72--dc21
                                                    97-28758
                                                    CIP

British Library Cataloguing-in-Publication Data
A catalogue record for this book is available from the British Library.

First published 1997
Reprinted 1998, 1999, 2003, 2005, 2006, 2008

Printed in Singapore.

PREFACE

The golden ratio and Fibonacci numbers have numerous applications which range from the description of plant growth and the crystallographic structure of certain solids to the development of computer algorithms for searching data bases. Although much has been written about these numbers, the present book will hopefully fill the gap between those sources which take a philosophical or even mystical approach and the formal mathematical texts. I have tried to stress not only fundamental properties of these numbers but their application to diverse fields of mathematics, computer science, physics and biology. I believe that this is the first book to take this approach since the application of models involving the golden ratio to the description of incommensurate structures and quasicrystals in the 1970's and 1980's.

This book will, hopefully, be of interest to the general reader with an interest in mathematics and its application to the physical and biological sciences. It may also be suitable supplementary reading for an introductory university course in number theory, geometry or general mathematics. Finally, the present volume should be sufficiently informative to provide a general introduction to the golden ratio and Fibonacci numbers for those researchers and graduate students who are working in fields where these numbers have found applications. Formal mathematics has been kept to a minimum, although readers should have a general knowledge of algebra, geometry and trigonometry at the high school or first year university level.

My own interest in the golden ratio and related topics developed from my involvement in research on the physical properties of incommensurate solids and quasicrystals. Over the years I have benefited greatly from discussions with colleagues in this field and many of the ideas presented in this book have been derived from these discussions. Without their involvement in my research in solid state physics, this book would not have been written. For their comments and ideas which eventually led to the present volume I would like to acknowledge Derek Lawther, Srinivas Veeturi, Dhiren Bahadur, Mike McHenry, Bob O'Handley and Bob March. I would also like to thank

v

Ewa Dunlap, Rene Coulombe, Jerry MacKay and Jody O'Brien for their advice and assistance during the preparation of the manuscript.

R.A. DUNLAP

Halifax, Nova Scotia
June 1997

CONTENTS

CHAPTER 1

INTRODUCTION

The *golden ratio* is an irrational number defined to be $(1+\sqrt{5})/2$. It has been of interest to mathematicians, physicists, philosophers, architects, artists and even musicians since antiquity. It has been called the *golden mean*, the *golden section*, the *golden cut*, the *divine proportion*, the *Fibonacci number* and the *mean of Phidias* and has a value of 1.61803... and is usually designated by the Greek character τ which is derived from the Greek word for *cut*. Although it is sometimes denoted ϕ, from the first letter of the name of the mathematician Phidias who studied its properties, it is more commonly referred to as τ while ϕ is used to denote $1/\tau$ or $-1/\tau$. The first known book devoted to the golden ratio is *De Divina Proportione* by Luca Pacioli [1445-1519]. This book, published in 1509, was illustrated by Leonardo da Vinci.

An irrational number is one which cannot be expressed as a ratio of finite integers. These numbers form an infinite set and some, such as π (the ratio of the circumference to the diameter of a circle) and e (the base of natural logarithms), are well known and have obvious applications in many fields. It is interesting to consider why the golden ratio has also attracted considerable attention and what its possible applications might be.

Certain irrational numbers can be expressed in the form

$$I = \frac{a+\sqrt{b}}{c} \tag{1.1}$$

where τ is defined for the values $a = 1$, $b = 5$ and $c = 2$. Other irrational numbers such as $a = 3$, $b = 3$, $c = 3$ would seem to have a more pleasing symmetry than the golden ratio and a similar value; 1.57735... . However, the golden ratio possesses a number of interesting and important properties which make it unique among the set of irrational numbers. Much has been written about the golden ratio and its applications in different fields (e.g. Brandmüller 1992, I. Hargittai 1992, Huntley

1

1990). While much of this work is scientifically valid and is based on the unique properties of τ as an irrational number, a significant portion of what has been written on τ is considerably more speculative. It is the intent of this book to provide scientifically valid information. However, a brief discussion of some of the more speculative claims concerning the golden ratio follows and this provides an interesting introduction to the remainder of this book.

Golden rectangles

The unique properties of the golden ratio were first considered in the context of dividing a line into two segments. If the line is divided so that the ratio of the total length to the length of the longer segment is the same as the ratio of the length of the longer segment to the length of the shorter segment then this ratio is the golden ratio. The so-called *golden rectangle* may be constructed from these line segments such that the length to width ratio (the aspect ratio, a) is τ. The ancient Greeks believed that a rectangle constructed in such a manner was the most aesthetically pleasing of all rectangles and they incorporated this shape into many of their architectural designs.

Figure 1.1 shows a number of rectangles with different aspect ratios. Although studies have shown that rectangles with a around 1.5 are more attractive to many people than those which are either more square (a near 1) or more elongated (large a), it is not obvious that the figure with $a = \tau$ is more aesthetically pleasing than those with a of $\sqrt{2}$, 3/2 or $\sqrt{3}$. It is, therefore, not clear that the particular properties of τ as an irrational number are of any fundamental importance to its role in the pleasing shape of the golden rectangle.

Art

The aesthetic appeal of the golden ratio in art has been the subject of a number of studies (e.g. Runion 1972). While it is true that many paintings include rectangular components which have aspect ratios near the golden ratio there is rarely any evidence that the artist considered the golden ratio in any conscious way in the composition of the painting. Rather it is likely that rectangular elements with aspect ratios near τ (or perhaps near $\sqrt{2}$, 3/2 or $\sqrt{3}$) provided pleasing proportions. In some cases artists have incorporated elements in their paintings which exhibit fivefold symmetry (see e.g. Dunlap 1992). As will be demonstrated in later chapters, there is a close relationship between the golden ratio and fivefold symmetry. In such cases the importance of the golden ratio in art is, perhaps,

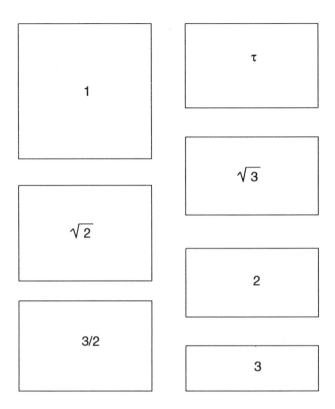

Fig. 1.1. Some rectangles with different aspect ratios, a (shown by the numbers inside the rectangles).

more definitive but certainly less direct.

The great pyramid

The relationship of the golden ratio to the design of the great pyramid of Cheops has also been the subject of some speculation (see Verheyen 1992). The great pyramid has a base edge length of about 230 m, a height of about 147 m (although about 9.5 m of this has weathered away) and an apex angle of approximately $\alpha = 63.43°$ (see Fig. 1.2). This apex angle is very close to the apex angle of the golden rhombus (63.435°) which has dimensions derived from the golden ratio and which is discussed in detail in Chapter 12. It has been suggested that the designers of the great pyramid were conscious of the relationship of the

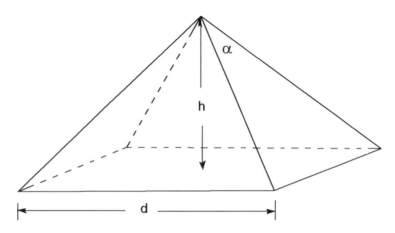

Fig. 1.2. Schematic representation of the great pyramid showing the height, h, the base dimension, d, where the circumference $c = 4d$ and the apex angle, α.

golden ratio and the pyramid's dimensions. A much older hypothesis has suggested that the dimensions of the great pyramid are related to the constant π. Specifically it has been speculated that the ratio of the circumference of the base of the pyramid, $c = 4d$, to its height, h, is 2π. It can be shown that this ratio is related to the apex angle of a pyramid by the expression

$$\frac{c}{h} = 8 \left(\frac{1}{2\cos\alpha} - \frac{1}{2} \right)^{1/2} . \tag{1.2}$$

A value of $c/h = 2\pi$ corresponds to an angle of $\alpha = 63.405°$. The difference between this angle and the apex angle of the golden rhombus results in a difference of about 22 *cm* in the edge length of the pyramid base. Since the base edges on the north and south sides of the great pyramid differ by 20 *cm* it would seem to be difficult to determine from the dimensions of the pyramid itself whether τ or π (if either) was a factor in its design; some insight into the philosophy of the pyramid designers would help to resolve this question. It may be that the similarity of certain aspects of the pyramid's geometry to either of these constants is purely coincidental. In fact some ratio of dimensions of virtually every object is likely to be close to some numerical constant of interest. Gillings (1972) has commented on the relevance of the golden ratio to the dimensions of the great pyramid as follows: "... the dimensions of the Eiffel Tower or Boulder Dam could be made to produce equally pretentious expressions of a mathematical connotation."

Pi

The relationship of the golden ratio to pi has generated some highly speculative hypotheses. The constant π is defined to be the ratio of the circumference, c, to the diameter, d, of a circle;

$$\pi = \frac{c}{d} \ .$$ (1.3)

The constant π has been studied in great detail by mathematicians and is one of the irrational numbers which has been calculated to the largest number of digits. It has been suggested (perhaps less than seriously) that c/d is not π but is a quantity related to the golden ratio. Two possible relationships (at least) have been suggested;

$$\frac{c}{d} = \frac{6\tau^2}{5}$$ (1.4)

and

$$\frac{c}{d} = \frac{4}{\sqrt{\tau}} \ .$$ (1.5)

These expressions yield values of 3.141641 ... and 3.144606 ..., respectively. These are close to the *accepted* value of π, 3.14159265 ... but are sufficiently different to preclude any serious consideration of these hypotheses.

Some of what has been said above about the golden ratio has some concrete evidence to support it; much more of it is highly speculative and some of it absolutely ludicrous. The possibility of the occurrence of τ in the dimensions of man-made objects (i.e. in art and architecture, for example), even when convincing evidence exists, is primarily of interest from an archeological, social or psychological point of view and does little to provide information concerning the interesting properties of τ itself. This book, therefore, deals primarily with the mathematical properties of τ as a unique irrational number and, to a large extent, its relationship to the series of so-called Fibonacci numbers. It is shown that the golden ratio plays a prominent role in the dimensions of all objects which exhibit fivefold symmetry. It is also shown that among the irrational numbers, the golden ratio is the most irrational and, as a result, has unique applications in number theory, search algorithms, the minimization of functions, network theory, the atomic structure of certain materials and the growth of biological organisms. The topics discussed in this book all deal with the unique properties of the golden ratio and the Fibonacci numbers and the applications of these mathematical concepts to topics which range from the methods of efficiently alphabetizing a list of names to the pattern of seeds in a sunflower.

CHAPTER 2

BASIC PROPERTIES OF THE GOLDEN RATIO

The golden ratio appears in some very fundamental relationships involving numbers from which many of its properties can be derived. One of the most basic occurrences of the golden ratio, and one of the most intriguing, involves the properties of numerical sequences. A numerical sequence is an ordered set of numbers which is generated by a well defined algorithm. One of the simplest methods of producing a numerical sequence is by the use of one or more seed values and an appropriate recursion relation.

One of the best known numerical sequences is the additive sequence. This is generated by the recursion relation

$$A_{n+2} = A_{n+1} + A_n \ . \tag{2.1}$$

That is, each term is equal to the sum of the two previous terms. This sequence requires two seed values, A_0 and A_1. The simple case of $A_0 = 0$ and $A_1 = 1$ may be considered as an example. This gives the sequence

$$0, 1, 1, 2, 3, 5, 8, 13, 21, 34, 55, 89, 144, \ \dots \ . \tag{2.2}$$

This sequence can be extended indefinitely by applying the recursion relation. It may also be extended to negative values of the index, n, by applying a recursion relation based on Eq. (2.1) to the values given in Eq. (2.2) yielding a sequence which extends indefinitely in both directions;

$$\dots \ 34, -21, 13, -8, 5, -3, 2, -1, 1, 0, 1, 1, 2, 3, 5, 8, \ \dots \ . \tag{2.3}$$

In this particular case the values of the terms with negative indices are numerically the same as the corresponding terms with positive indices but they alternate in sign. This is an interesting property of this particular additive sequence which will be discussed further in Chapter 5, although it is not a property of additive sequences in general.

Another simple numerical sequence, referred to as the geometric sequence, is generated by the recursion relation

$$A_{n+1} = \alpha A_n \ .$$
 (2.4)

That is, each term is the previous term multiplied by some constant factor. This sequence may be generated on the basis of one seed value and the value of the constant factor. A simple example uses $A_0 = 1$ and $\alpha = 2$. This gives the familiar sequence of powers of 2;

$$1, \ 2, \ 4, \ 8, \ 16, \ 32, \ 64, \ 128, \ 256, \ 512, \ \dots \ .$$
 (2.5)

Again it is straightforward to extend this sequence to negative values of the indices;

$$\dots \ \frac{1}{32}, \ \frac{1}{16}, \ \frac{1}{8}, \ \frac{1}{4}, \ \frac{1}{2}, \ 1, \ 2, \ 4, \ \dots \ .$$
 (2.6)

A comparison of Eqs. (2.3) and (2.6) would seem to illustrate the fundamental differences between additive and geometric sequences. However, these differences are the result of the particular choice of the multiplicative constant in Eq. (2.4). Different choices for this quantity can yield very different results. Consider, for example, the possibility that a sequence could be both additive and geometric; that is, the terms would satisfy both Eq. (2.1) and Eq. (2.4). These two equations can be combined to give the constraining relations for α. From Eq. (2.4) we can write

$$A_{n+2} = \alpha A_{n+1} = \alpha^2 A_n \ .$$
 (2.7)

Eqs. (2.1) and (2.7) yield the relation

$$\alpha^2 A_n = \alpha A_n + A_n$$
 (2.8)

or simply

$$\alpha^2 - \alpha - 1 = 0 \ .$$
 (2.9)

This equation is known as the Fibonacci quadratic equation and is easily solved to yield the two roots

$$\alpha_1 = \frac{1+\sqrt{5}}{2} = \tau \qquad (2.10)$$

and

$$\alpha_2 = \frac{1-\sqrt{5}}{2} = -\frac{1}{\tau} . \qquad (2.11)$$

It is straightforward to construct a geometric sequence using the value of α_1 as the constant factor and a seed value of (say) $A_0 = 1$. This gives

$$1, \ \tau, \ \tau^2, \tau^3, \tau^4, \tau^5, \ \dots \ . \qquad (2.12)$$

Extending this to negative indices yields

$$\dots \ \tau^{-3}, \tau^{-2}, \tau^{-1}, \ 1, \ \tau, \ \tau^2, \tau^3, \ \dots \ . \qquad (2.13)$$

Using the seed values of $A_0 = 1$ and $A_1 = \tau$ from Eq. (2.12) a corresponding additive sequence may be constructed using the recursion relation of Eq. (2.1). For negative and positive indices this sequence is

$$\dots \ -3\tau +5, \ 2\tau -3, \ -\tau +2, \ \tau -1, \ 1, \ \tau, \ \tau +1, \ 2\tau +1, \ 3\tau +2, \ \dots \ . \qquad (2.14)$$

Numerically the terms in this sequence are the same as those in the geometric sequence in Eq. (2.13). These terms may be equated to yield some interesting relationships between powers of τ and linear expressions in τ. Some of these are

$$2\tau - 3 = \tau^{-3}$$
$$-\tau + 2 = \tau^{-2}$$
$$\tau - 1 = \tau^{-1}$$
$$1 = 1$$
$$\tau = \tau$$
$$\tau + 1 = \tau^2$$
$$2\tau + 1 = \tau^3 . \qquad (2.15)$$

In general, powers of the golden ratio may be expressed as

Table 2.1. Some coefficients and exponents in the relationship given by Eq. (2.16).

n	a_n	a_{n-1}
-8	-21	34
-7	13	-21
-6	-8	13
-5	5	-8
-4	-3	5
-3	2	-3
-2	-1	2
-1	1	-1
0	0	1
1	1	0
2	1	1
3	2	1
4	3	2
5	5	3
6	8	5
7	13	8
8	21	13

$$a_n \tau + a_{n-1} = \tau^n \qquad (2.16)$$

where the coefficients, a_n, as given in Table 2.1 are the A_n of the additive sequence in Eq. (2.3). This relationship is discussed further in Chapter 5.

Another sequence which is both additive and geometric can be derived using the other root of the quadratic equation as given by Eqs. (2.11) and (2.15), $\alpha_2 = -\tau^{-1} = 1 - \tau$. This gives the sequence

$$\ldots \; -\tau^3, \tau^2, -\tau, \; 1, \; -\tau^{-1}, \tau^{-2}, -\tau^{-3} \quad \ldots \qquad (2.17)$$

and the corresponding sequence based on the additive recursion relation is found to be

$$\ldots -3-\frac{2}{\tau}, \; 2+\frac{1}{\tau}, \; -1-\frac{1}{\tau}, \; 1, \; -\frac{1}{\tau}, \; 1-\frac{1}{\tau}, \; 1-\frac{2}{\tau}, \; 2-\frac{3}{\tau}, \ldots \; . \qquad (2.18)$$

Equating terms from Eqs. (2.17) and (2.18) allows for the derivation of relations of the form

$$a_{n+1} + \frac{a_n}{\tau} = \tau^n \qquad (2.19)$$

where the coefficients are again the terms in the additive sequence of Eq. (2.3). It can be shown that these expressions are algebraically equivalent to those of Eq. (2.16) by multiplying both sides of Eq. (2.19) by τ.

The above discussion concerning numerical sequences illustrates the relationship of the golden ratio to some fundamental properties of numbers. Additional insight into the properties of the golden ratio may be gained by taking a somewhat more geometric approach. In fact, it is this occurrence of the golden ratio which is responsible for its appeal to the ancient philosophers and for the derivation of its name; the golden *ratio*. Consider a line AC which is divided by a point B as illustrated in Fig. 2.1 in such a way that the ratio of the lengths of the two segments is the same as the ratio of the length of the longer segment to the entire line. If the length AB is arbitrarily set equal to 1 and the length of the total line is called x then the segment $BC = x-1$ and the ratios of lengths may be expressed as

$$\frac{x}{1} = \frac{1}{x-1} \qquad (2.20)$$

or

$$x^2 - x - 1 = 0 \ . \qquad (2.21)$$

This is the Fibonacci equation which has the roots given in terms of the golden ratio by Eqs. (2.10) and (2.11); τ and $-1/\tau$. Obviously it is the positive root which has some physical significance in the context of this problem. Alternately the total length of the line may be set to 1 and segment AB may be arbitrarily called x. The ratios are then

$$\frac{1}{x} = \frac{x}{1-x} \qquad (2.22)$$

Fig. 2.1. Sectioning of the line.

or

$$x^2 + x - 1 = 0 \ . \tag{2.23}$$

This quadratic equation has roots which may be expressed in terms of the golden ratio as

$$x_1 = \frac{\sqrt{5}-1}{2} = \frac{1}{\tau} \tag{2.24}$$

and

$$x_2 = -\frac{\sqrt{5}+1}{2} = -\tau \ . \tag{2.25}$$

Again only the positive root has physical significance and shows the ratio of lengths to be related to the golden ratio.

Some interesting mathematical relationships involving the golden ratio can be derived by combining powers of τ. For example, a simple inspection of relationships such as those shown in Eq. (2.15) and Table 2.1, will allow for the derivation of expressions involving both positive and negative powers of the golden ratio. The simplest of these is

$$\tau^n + (-1)^n \tau^{-n} = L_n \tag{2.26}$$

where L_n is an integer that takes on values L_n=1, 3, 4, 7, 11, 18,... for n=1, 2, 3, 4, 5, 6,... . These are the so-called Lucas numbers and are disussed further in Chapter 6. This expression is somewhat remarkable as it shows that the sum of two irrational numbers can be ᴗqual to a rational number.

Another interesting relationship involving the golden ratio may be obtained directly from the Fibonacci quadratic equation, Eq. (2.9). This may be written for τ as

$$\tau = \sqrt{1+\tau} \ . \tag{2.27}$$

Substituting the left hand side for τ in the square root on the right hand side gives

$$\tau = \sqrt{1+\sqrt{1+\tau}} \ . \tag{2.28}$$

This procedure may be continued indefinitely to yield

$$\tau = \sqrt{1+\sqrt{1+\sqrt{1+\sqrt{1+\sqrt{1+...}}}}} \ . \tag{2.29}$$

Along similar lines it is known that the positive root of Eq. (2.23) is $1/\tau$. This expression may be rearranged and the substitution for the term in the square root performed indefinitely to give

$$\frac{1}{\tau} = \sqrt{1 - \sqrt{1 - \sqrt{1 - \sqrt{1 - \sqrt{1 - \dots}}}}} \qquad . \tag{2.30}$$

The expression in Eq. (2.30) provides one means of calculating the golden ratio to a high degree of accuracy using a computer. It is, however, less time consuming to calculate τ directly on the basis of Eq. (2.10) by first calculating the square root of 5. An irrational square root can be calculated to an arbitrary accuracy using a simple iterative technique. To calculate a square root to an accuracy of N digits requires a number of basic arithmetic operations which is proportional to N^2. An early report of the use of a computer to calculate the golden ratio to high accuracy provided τ to 4599 decimal places; see Berg (1966). This required about 20 minutes on an IBM 1401 main frame computer. Today this calculation can be done on an IBM Pentium personal computer in about 2 seconds. It is straightforward to determine the validity of the calculated values. One method is to substitute the calculated value of τ into the Fibonacci equation (Eq. (2.9)) and perform the operations to the required number of decimal places and show that the identity holds. An equivalent method is to calculate the reciprocal of τ and show that $1/\tau = \tau - 1$ holds to the required accuracy.

CHAPTER 3

GEOMETRIC PROBLEMS IN TWO DIMENSIONS

The golden ratio plays an important role in the dimensions of many geometric figures, in both two and three dimensions. The simplest appearance of τ in geometry occurs in two dimensional figures, and it is here that the close affinity of the golden ratio with fivefold symmetry is first apparent. Among the most aesthetically appealing two dimensional shapes are the regular polygons. These are figures which have all edges equal and all interior angles equal and less than 180° (see e.g. Dunlap 1997). The simplest of these is the equilateral triangle, with three edges. In general a regular n-gon has n edges and interior angles which are given by the relation

$$\alpha = [1 - \frac{2}{n}] \cdot 180° \quad . \tag{3.1}$$

It is the regular pentagon, with $n = 5$, which exhibits fivefold symmetry in two dimensions, and Eq. (3.1) gives the interior angle of $\alpha = 108°$. This is illustrated in Fig. 3.1a. If the edge lengths of the pentagon are 1 then it can be shown that the diagonal as illustrated in Fig. 3.1b has a length τ. The pentagon may be divided into three isosceles triangles as shown in Fig. 3.1c by two diagonals with one vertex in common. Two of the triangles are obtuse with edge lengths $1 : \tau : 1$ and one is acute with edge lengths $\tau : 1 : \tau$. These are usually referred to as the golden gnomons and the golden triangle, respectively. The angle relationships as shown in the figure can be expressed as

$$\beta + 2\gamma = \alpha$$

$$\delta + \gamma = \alpha$$

$$\alpha + 2\gamma = \beta + 2\delta = 180° \quad . \tag{3.2}$$

These equations may be combined with Eq. (3.1) to obtain values of the angles $\beta = \gamma = 36°$ and $\delta = 72°$. Figure 3.1 illustrates that the ability to construct two line segments with length ratios of $1 : \tau$ provides a simple means of constructing a regular pentagon. Ancient mathematicians and philosophers showed interest in

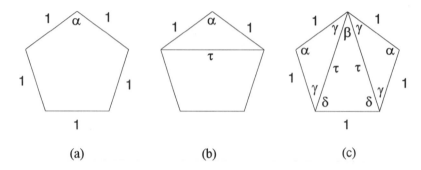

Fig. 3.1. (a) The regular pentagon with an edge length of 1, (b) the regular pentagon showing the diagonal of length τ and (c) the regular pentagon dissected into two golden triangles and a golden gnomon.

this problem because of the relevance of the pentagon in various aspects of art, architecture and religion. A straightforward method of constructing a line segment of length τ is shown in Fig. 3.2. This construction can be extended to produce a regular pentagon and the details of this construction are described in Appendix I.

Figure 3.3 shows a unique property of the golden gnomon and the golden triangle. Each may be dissected into two smaller triangles, one of which is a golden gnomon and the other is a golden triangle. This characteristic is the direct result of the fact that the golden ratio satisfies the relationship

$$\frac{1}{\tau} = \tau - 1 \quad . \tag{3.3}$$

This dissection procedure which may be referred to as inflation (as it increases the number of triangles) can be continued indefinitely, dividing resulting gnomons and triangles into smaller and smaller gnomons and triangles. An analogous procedure, which may be referred to as deflation, combines a triangle and a gnomon into a larger triangle or gnomon. These concepts will be discussed further in later chapters with reference to Fibonacci sequences and quasicrystals. It can be readily seen from an inspection of Fig. 3.3 that the inflation of one golden gnomon to a smaller golden gnomon represents a reduction in the linear dimensions of the gnomon by a factor of τ and a reduction in the area of the gnomon by a factor of τ^2. This same relationship holds for the inflation of the golden triangle as well.

The angles involved in the regular pentagon and in the golden gnomon and triangle are all multiples of $360°/10 = 36°$. Since the linear dimensions of the triangles involving these angles are related to the golden ratio, it is apparent that the trigonometric functions of angles related to $36°$ should also be related to the golden ratio. Table 3.1 gives trigonometric functions for some of these angles.

Another two dimensional figure which is closely related to the golden ratio is the golden rectangle as shown in Fig. 3.4. This rectangle has edge lengths which are in the ratio τ. The aesthetic appeal of the golden rectangle has been the subject of considerable discussion as has its role in art and Hellenic architecture. It has a number of interesting mathematical properties, and some of these will be discussed here and in the next chapter. The golden rectangle may be divided into a square and a smaller golden rectangle as shown in Fig. 3.4. This inflation reduces the linear dimension of the golden rectangle by a factor of τ and the area by a factor of τ^2. It is, therefore, a process which is analogous to the procedure for golden triangles and is a process which can be repeated indefinitely. Figure 3.5 shows several inflations of the golden rectangle. The relationship of powers of τ as given by Eq. (2.15) can be seen from the geometric analysis of Fig. 3.5 as the progressive

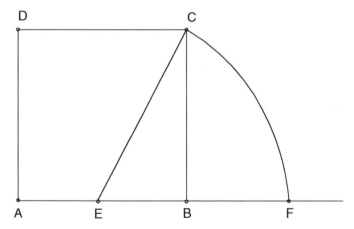

Fig. 3.2. Construction of a line segment with length τ. A square *ABCD* is constructed. The midpoint of the base of the square (point E) is located and a compass is used to draw an arc through point C with its center at point E. This arc intercepts the extension of the baseline of the square at point F. The ratio of lengths *AF* to *AB* is the golden ratio. A simple geometric calculation will show this to be true.

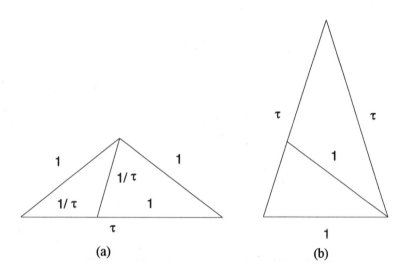

Fig. 3.3. (a) The golden gnomon and (b) the golden triangle. The dissection into smaller golden gnomons and triangles is illustrated.

Table 3.1. Trigonometric functions related to the golden ratio.

angle(θ)	sin (θ)	cos (θ)
18°	$\dfrac{1}{2}\sqrt{1-\dfrac{1}{\tau}}$	$\dfrac{1}{2}\sqrt{2+\tau}$
36°	$\dfrac{1}{2}\sqrt{2-\dfrac{1}{\tau}}$	$\dfrac{1}{2}\sqrt{1+\tau}$
54°	$\dfrac{1}{2}\sqrt{1+\tau}$	$\dfrac{1}{2}\sqrt{2-\dfrac{1}{\tau}}$
72°	$\dfrac{1}{2}\sqrt{2+\tau}$	$\dfrac{1}{2}\sqrt{1-\dfrac{1}{\tau}}$

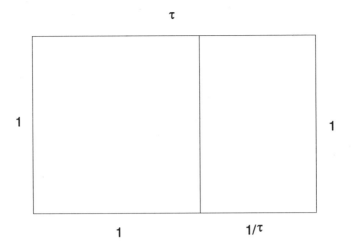

Fig. 3.4. The golden rectangle dissected into a square and a smaller golden rectangle.

inflations by a factor of τ yield golden rectangles with longest edge lengths which follow the sequence

$$\tau, \ 1, \ \tau-1, \ -\tau+2, \ 2\tau-3, \ -3\tau+5, \ \ldots \tag{3.4}$$

which is the sequence of Eq. (2.14) with decreasing indices.

Figure 3.5 also shows another interesting property; The diagonal of the original rectangle is perpendicular to the diagonal of the smaller rectangle. These diagonals are also the diagonals of alternating golden rectangles in the inflation process. This means that the inflation of the golden rectangles will converge at the point given by the intersection of the diagonals. Connecting vertices of this progression of golden rectangles with suitably curved lines as illustrated in Fig. 3.6, will yield a spiral which converges at the intersection of the two diagonals. This same spiral can be constructed by connecting the acute vertices of the golden triangles in a progression of inflated triangles as shown in Fig. 3.7. This particular spiral is referred to as the equiangular or logarithmic spiral and is given by the polar equation

$$r = r_0 e^{\theta \cot \alpha} \ . \tag{3.5}$$

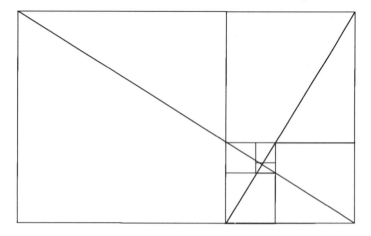

Fig. 3.5. Inflation of the golden rectangle showing the perpendicular diagonals.

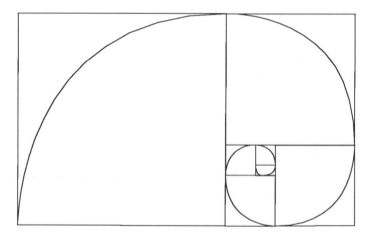

Fig. 3.6. Construction of the equiangular or logarithmic spiral from a sequence of golden rectangles.

Here the radius of the spiral, r, as measured from the pole, or point of intersection of the two diagonals, is expressed as a function of the angle θ. The quantity r_0 is a constant related to the overall dimensions of the spiral and the quantity α is a constant for a given spiral and is a measure of how tightly the spiral is wound. In the limiting case $\alpha = 90°$ and $\cot \alpha = 0$. Equation (3.5) then reduces to $r = r_0$

which is the polar equation of a circle with radius r_0. The logarithmic spiral plays an important role in the growth and structure of certain biological systems and will be discussed further in Chapter 13.

The discussion in this chapter has shown that certain two dimensional figures can be inflated indefinitely according to algorithms which have their basis in the mathematical properties of the golden ratio. Similarly an inverse process known as deflation of the logarithmic spiral and two and three dimensional tilings as will be discussed in later chapters. As well, it has been seen that the golden ratio appears in the dimensions of figures which exhibit fivefold symmetry. This feature is even more prevalent in three dimensional geometry and is discussed further in the next chapter.

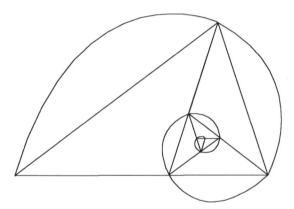

Fig. 3.7. Construction of the equiangular or logarithmic spiral from a sequence of golden triangles.

CHAPTER 4

GEOMETRIC PROBLEMS IN THREE DIMENSIONS

In two dimensions the number of regular n-gons with interior angles defined by Eq. (3.1) is infinite. In three dimensions the regular polyhedra may be defined in an analogous way; All faces are the same regular polygon and each vertex is convex (as viewed from outside the figure). There are precisely five such figures and these are known as the Platonic solids. It is, perhaps, curious that only five such solids exist but a simple proof of this fact may be given in the following manner. Each face of the regular polyhedron is a regular polygon with n edges. From the discussion in the previous chapter it is known that values of n which are permitted are the integers

$$3 \le n < \infty \tag{4.1}$$

with the interior angles, α, related to n by Eq. (3.1). Each vertex of the three dimensional polygon is defined by the intersection of a number of faces, m. In order to form a vertex the integer m is constrained by

$$m \ge 3 \tag{4.2}$$

(If $m = 2$ then an edge, not a vertex, is formed). In order for a convex vertex to be formed it is also necessary that

$$m\alpha < 360° \quad . \tag{4.3}$$

If $m\alpha = 360°$ then the vertex is merely a point on a plane and if $m\alpha > 360°$ then the faces overlap. The conditions as described in Eq. (3.1) and Eqs. (4.1) through (4.3) allow for the determination of the values of n and m for permissible regular polyhedra. There are only five combinations of integer values of n and m which satisfy these equations and these combinations are listed in Table 4.1. These correspond to the five Platonic solids as shown in Fig. 4.1. A relationship, known as Euler's formula, exists between the values of e, f and v, the number of edges,

Table 4.1. Characteristics of the five Platonic solids. The quantities n and m are the number of edges per face and the number of faces per vertex, respectively. The quantities e, f and v are the total number of edges, faces and vertices for the solid.

solid	n	m	e	f	v
tetrahedron	3	3	6	4	4
cube (hexahedron)	4	3	12	6	8
octahedron	3	4	12	8	6
dodecahedron	5	3	30	12	20
icosahedron	3	5	30	20	12

faces and vertices of the polyhedron, respectively, in the table;

$$f + v = e + 2 \ . \tag{4.4}$$

This equation applies to all convex polyhedra, not only the Platonic solids. A large number of convex polyhedra which are not regular exist. One common example of these is the traditional design of a soccer ball which consists of 12 pentagonal and 20 hexagonal faces.

As the previous chapter indicated, the golden ratio is of relevance to the geometry of figures with fivefold symmetry and it is the dodecahedron and the icosahedron which are of particular interest to the present discussion. If these two Platonic solids are constructed with an edge length of one unit, then the total surface areas and volumes of the solids are given in Table 4.2. It is obvious from these quantities that the golden ratio plays an important role in the dimensions of these solids.

The importance of the golden ratio is also apparent in the relationship of the icosahedron to the golden rectangle. Three golden rectangles may be arranged so that they are mutually perpendicular and their centers are coincident. The twelve vertices (four vertices for each of the three rectangles) lie at the vertices of an icosahedron as illustrated in Fig. 4.2. If the golden rectangles have dimensions 1 by τ then the resulting icosahedron has an edge length of 1.

Certain relationships are apparent between the values of n, m, e, f and v for some of the Platonic solids. Specifically, the cube and the octahedron, as well as the dodecahedron and the icosahedron, have the same values of e, while values of n and m, as well as f and v are interchanged. Solids which are related by the same

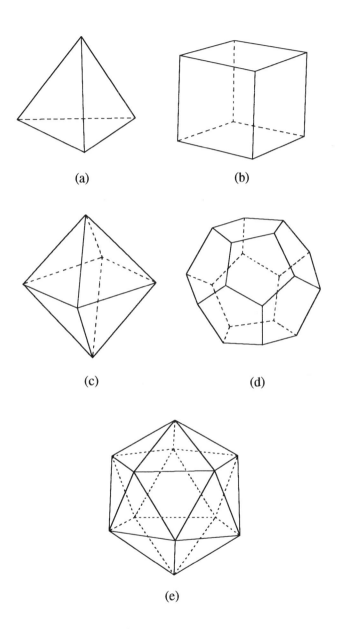

Fig. 4.1. The five Platonic solids: (a) tetrahedron, (b) cube (or hexahedron), (c) octahedron, (d) dodecahedron and (e) icosahedron.

Table 4.2. Surface areas and volumes of the dodecahedron and the icosahedron with edge lengths of 1.

solid	surface	volume
dodecahedron	$\dfrac{15\tau}{\sqrt{3-\tau}}$	$\dfrac{5\tau^3}{6-2\tau}$
icosahedron	$5\sqrt{3}$	$\dfrac{5\tau^5}{6}$

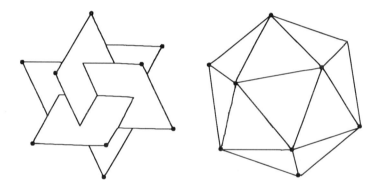

Fig. 4.2. Construction of an icosahedron from three mutually perpendicular golden rectangles.

values of e are sometimes referred to as duals. The tetrahedron is unique as it is not paired with any of the other solids. It is, therefore, said to be self-dual. These equalities between certain geometric factors of the Platonic solids result from similarities in their symmetry and allow for the mapping of one solid into another. This is, perhaps, easiest to visualize for the cube and the tetrahedron. The cube has six faces while the octahedron has six vertices. If a vertex is constructed at the center of each face of the cube and these vertices are connected together by edges, an octahedron is formed as shown in Fig. 4.3. Similarly, the octahedron has eight faces and the cube has eight vertices. Constructing a vertex at the center of each face of the octahedron will produce a cube. Repeating this process produces smaller and smaller cubes and octahedra which are rescaled by a constant factor.

Another method of mapping a cube into an octahedron is related to the fact that both solids have 12 edges. If perpendicular bisectors of each of the twelve edges of the cube are constructed these will form the edges of an octahedron as seen in Fig. 4.4. The reverse procedure of constructing perpendicular bisectors of the edges of an octahedron to form a cube is also valid. The rescaling of the edge length of the solid by each of these mapping procedures, face-to-vertex and edge-to-edge, are given in Table 4.3. Perhaps not surprisingly, the factor $\sqrt{2}$ appears in relationships between the cube and the octahedron.

The same procedures can be used to map a dodecahedron into an icosahedron and vice versa as both solids have 30 edges and the dodecahedron has 12 faces and 20 edges while these values are interchanged for the icosahedron. An example of the edge-to-edge mapping between a dodecahedron and an icosahedron is illustrated in Fig. 4.5. Values of the edge length ratio obtained during this mapping procedure are given in Table 4.3. It is seen that the golden ratio appears frequently in these relationships. Applying either of these two mapping methods to the tetrahedron will produce another tetrahedron with dimensions as given in the table.

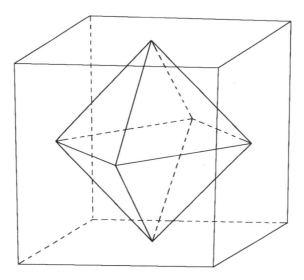

Fig. 4.3. Vertex-to-face relationship between the cube and the octahedron.

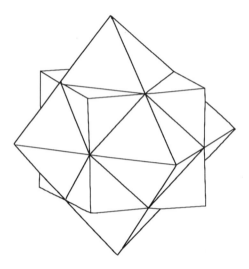

Fig. 4.4. Edge-to-edge relationship between the cube and the octahedron.

Curiously the value of $f = 12$ for the dodecahedron and $v = 12$ for the icosahedron is the same as the value of e for both the cube and the octahedron. It might be hypothesized that placing a vertex at the center of each edge of (say) an octahedron would yield an icosahedron. This hypothesis, however, is incorrect. If, on the other hand, each edge of an octahedron is divided into two segments with relative lengths in the ratio of $1 : \tau$ then these points do form the vertices of an icosahedron. Some care is required in locating these vertices. Four edges form each vertex of the octahedron. Two opposite edges are divided so that the longer edge segment is adjacent to the vertex while the other two opposite edges are divided so that the shorter edge segment is adjacent to the vertex. Each vertex may be treated in this manner.

In the case of two dimensional figures it is possible to relax some of the angular relationships which were applied to obtain the regular n-gons while keeping all edge lengths equal. Requiring that all interior angles be less than 180° but not requiring them to be equal will produce figures such as a variety of rhombuses (see Fig. 4.6a). Some of these are of particular relevance to the golden ratio and will be discussed further in Chapters 11 and 12. Allowing some interior angles to be greater than 180° but requiring that all acute angles are equal and all

Table 4.3. Edge length ratios for Platonic solids produced by face-to-vertex mapping and edge-to-edge mapping. From Dunlap (1990).

	edge length ratios	
solids	edge-to-edge	face-to-vertex
tetrahedron-tetrahedron	1 : 1	1 : 1/2
cube-octahedron	1 : $\sqrt{2}$	1 : $1/\sqrt{2}$
dodecahedron-icosahedron	1 : τ	1 : $\tau^2/\sqrt{5}$

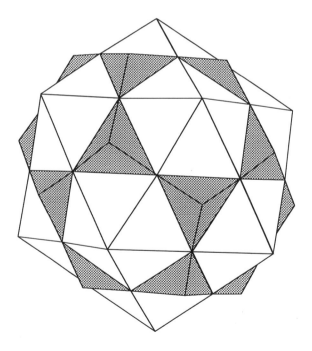

Fig. 4.5. Edge-to-edge relationship between the dodecahedron and the icosahedron.

obtuse angles are equal will yield figures such as the pentagram or five pointed star shown in Fig. 4.6b. This figure has obvious relationships to the golden ratio as it may be constructed from the five diagonals of a regular pentagon.

A similar relaxing of some of the criteria for the construction of the regular polyhedra in three dimensions will yield additional solids. If the requirement that all vertices be convex (as viewed from the outside) is eliminated and faces are allowed to be *regular n*-gons of the type shown in Fig. 4.6b then precisely four additional *regular* polyhedra are existent. The requirement that all faces are identical and that all convex vertices are equivalent is imposed. Johannes Kepler

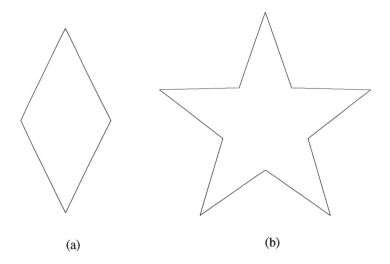

(a)   (b)

Fig. 4.6. (a) A rhombus and (b) the pentagram or five pointed star.

Table 4.4. Properties of the Kepler-Poinsot solids. The quantity v refers to a convex vertex.

solid	e	f	v
small stellated dodecahedron	30	12	12
great dodecahedron	30	12	12
great stellated dodecahedron	30	12	20
great icosahedron	30	20	12

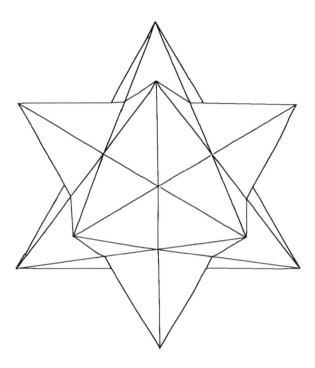

Fig. 4.7. The small stellated dodecahedron.

discovered two of these four additional solids. These are the small stellated dodecahedron and the great stellated dodecahedron and are illustrated in Figs. 4.7 and 4.8, respectively. Both these solids are formed from twelve pentagram faces. In the case of the small stellated dodecahedron the vertices of the pentagrams are arranged so that the intersection of five pentagrams forms a vertex of the solid. It may be viewed as a dodecahedron with a pentagonal pyramid on each face. The great stellated dodecahedron is formed in a similar manner except that the vertices are formed from the vertices of three pentagrams. The characteristics of these solids are given in Table 4.4.

Two centuries after Kepler described the stellated dodecahedra Louis Poinsot described two additional regular solids (see e.g. Holden 1971). The great dodecahedron is formed from twelve faces which are regular pentagons and is shown in Fig. 4.9. The great icosahedron is formed from twenty faces which are equilateral triangles and is shown in Fig. 4.10. The characteristics of these solids are described in Table 4.4. The great icosahedron is a particularly interesting figure as it demonstrates that twenty equilateral triangles can be arranged to form regular solids with thirty edges and twelve vertices in two distinctly different ways.

An inspection of Table 4.4 will demonstrate the relationship of the Keplerian and Poinsot solids. The small stellated dodecahedron and the great dodecahedron are duals and the great stellated dodecahedron and the great icosahedron are duals. The mapping relationships between vertices and faces and between edges and edges as they have been described above for the Platonic solids can be applied to the Kepler-Poinsot duals as well. As a result of the obvious fivefold symmetry of these four new solids the scaling relations for the edge lengths in the mapping transformations are related to the golden ratio, as are numerous ratios of dimensions of the solids themselves.

Fig. 4.8. The great stellated dodecahedron.

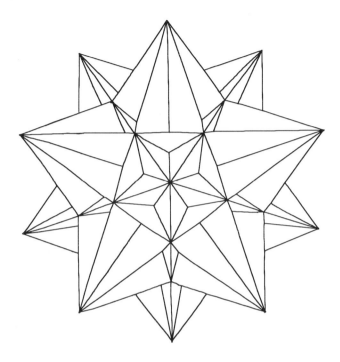

Fig. 4.9. The great dodecahedron.

Along the lines of the proof given above for the existence of five Platonic solids, it can also be shown that the Kepler-Poinsot solids constitute the complete set of regular polyhedra which allow for convex vertices. Thus only nine regular polyhedra can exist, which consist of four sets of duals and the tetrahedron which is self-dual.

The present chapter has demonstrated that the solids which exhibit fivefold symmetry (which constitute six of the nine solids) have linear dimensions, surface areas and volumes which are related to the golden ratio. It is also shown that the three sets of fivefold symmetry duals can be rescaled by factors involving the golden ratio by utilizing face-to-vertex or edge-to-edge mapping transformations. This is similar to the deflation operations which were seen previously for golden triangles and golden rectangles and will play an important role in the discussion of tilings in Chapters 11 and 12.

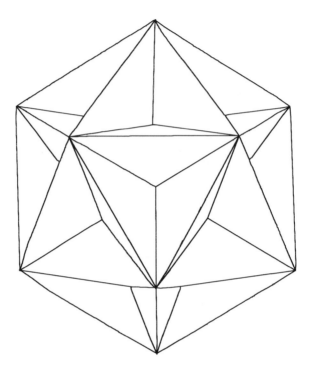

Fig. 4.10. The great icosahedron.

CHAPTER 5

FIBONACCI NUMBERS

The Italian mathematician Leonardo de Pisa was born in Pisa around 1175 AD. He is commonly known as Fibonacci which is a shortened form of Filius Bonaccio (son of Bonaccio). His father, Bonaccio, was a customs inspector in the city of Bugia on the north coast of Africa (presently Bougie in Algeria) and as a result, Fibonacci was educated by the Mohammedans of Barbary. He was taught the Arabic system of numbers and in the early thirteenth century returned to Italy to publish the book *Liber Abaci* (*Book of the Abacus*) in 1202 (Leonardo di Pisa 1857). This book introduced the Arabic system of numbers to Europe and is responsible for Fibonacci's reputation as the most accomplished mathematician of the middle ages. The book also posed a problem involving the progeny of a single pair of rabbits which is the basis of the Fibonacci sequence (or Fibonacci series). It was, however, Edouard Lucas, whose contribution to this area of mathematics will be discussed in detail in the next chapter, who *rediscovered* the Fibonacci sequence in the late nineteenth century, and properly attributed it to its original founder.

The rabbit problem is as follows:

A pair of adult rabbits produces a pair of baby rabbits once each month. Each pair of baby rabbits requires one month to grow to be adults and subsequently produces one pair of baby rabbits each month thereafter. Determine the number of pairs of adult and baby rabbits after some number of months. It is also assumed that rabbits are immortal.

This problem may be expressed mathematically in this way: The number of adult rabbit pairs in a particular month (say month $n+2$), A_{n+2}, is given by the number of adult rabbit pairs in the previous month, A_{n+1}, plus the number of baby rabbit pairs from the previous month which grow to be adults, b_{n+1};

$$A_{n+2} = A_{n+1} + b_{n+1} \ . \tag{5.1}$$

In a given month (say month $n+1$), the number of pairs of baby rabbits will be equal to the number of adult rabbit pairs in the previous month;

$$b_{n+1} = A_n \quad . \tag{5.2}$$

Combining Eqs. (5.1) and (5.2) gives the recursion relation for the number of adult rabbit pairs as

$$A_{n+2} = A_{n+1} + A_n \quad . \tag{5.3}$$

This recursion relation is identical to the expression for the additive sequence given by Eq. (2.1) and shows that the number of adult rabbit pairs will follow this kind of sequence. From Eq. (5.2) it is easy to see that the number of baby rabbit pairs will also follow the same sequence but will be displaced by one month. Since the total number of rabbit pairs is equal to the number of adult rabbit pairs plus the number of baby rabbit pairs then this quantity will also follow an additive sequence.

Table 5.1. Number of baby rabbit pairs, b_n , the number of adult rabbit pairs, A_n , and the number of total rabbit pairs, $(b+A)_n$ as a function of the number of months.

month	b_n	A_n	$(b+A)_n$
1	0	1	1
2	1	1	2
3	1	2	3
4	2	3	5
5	3	5	8
6	5	8	13
7	8	13	21
8	13	21	34
9	21	34	55
10	34	55	89
11	55	89	144
12	89	144	233
13	144	233	377
14	233	377	610
15	377	610	987

As an example the simple case of one adult rabbit pair which produces a pair of baby rabbits in the second month may be considered. Table 5.1 shows the population of rabbits as a function of the number of months. In general the table shows that

$$(b + A)_n = A_{n+1} = b_{n+2} \ .$$ (5.4)

Each of these sequences follows the additive sequence given in Eq. (2.2). The numbers which form this sequence are known as the Fibonacci numbers (see Vorobyov 1963), F_n, where

$$F_n = A_n \ .$$ (5.5)

Thus beginning with the seed values which represent the number of adult and baby rabbit pairs in the first month the Fibonacci numbers may be calculated for all values of the index n, as given in Eq. (2.3). Although in the context of the rabbit problem Fibonacci numbers with negative indices have no physical meaning, they are important for some applications. Appendix II gives values of the Fibonacci numbers for indices from 0 to 100.

There are numerous occurrences of the Fibonacci numbers in problems related to a number of diverse fields. A few of the more interesting ones are described here.

Along the lines of the rabbit breeding problem, the genealogy of bees is described in terms of Fibonacci numbers. The male bee, or drone, hatches from an egg which has not been fertilized. Fertilized eggs produce only females which become either workers or queens. Thus, the family tree of a single male bee may be constructed as shown in Fig. 5.1. The number of male bees and the number of female bees as well as the total number of bees is tabulated in Table 5.2. These are seen to follow the sequence of Fibonacci numbers and the recursion relations as derived above for the number of rabbit pairs can be shown to be applicable to the bee problem as long as it is assumed that bees, like rabbits, are immortal.

Fibonacci numbers also appear in the field of optics. A system is constructed from two plane sheets of glass with slightly different indices of refraction. Rays of light which are incident on one piece of glass will undergo various numbers of internal reflections before emerging. Some examples of possible ray paths in this system are illustrated in Fig. 5.2. In each case the number of emergent beams, B_n, for n internal reflections is equal to a Fibonacci number. It is easy to see from an inspection of the figure that

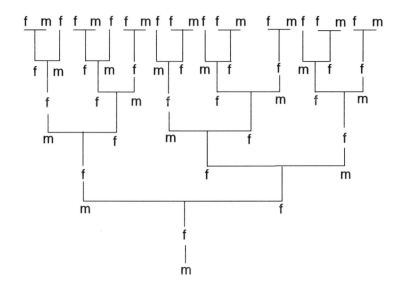

Fig. 5.1. Genealogy of a male (drone) bee; m = male, f = female.

Table 5.2. Genealogy of a male (drone) bee.

generation	n_{males}	$n_{females}$	n_{total}
1	1	0	1
2	0	1	1
3	1	1	2
4	1	2	3
5	2	3	5
6	3	5	8
7	5	8	13
8	8	13	21
9	13	21	34
10	21	34	55

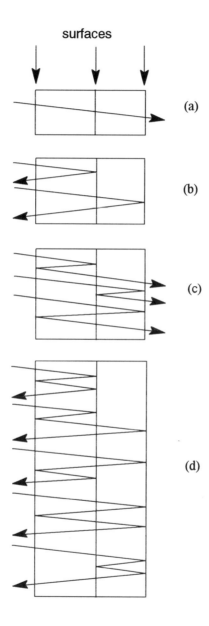

Fig. 5.2. Internal reflections for a beam of light incident upon two sheets of glass. Possible ray paths are shown for (a) zero internal reflections, (b) one internal reflection, (c) two internal reflections and (d) three internal reflections.

$$B_n = F_{n+2} \quad .$$
(5.6)

As a final example of the occurrence of Fibonacci numbers, a somewhat more mathematical problem will be considered here. A staircase consists of n stairs. This is climbed by taking either one step or two steps at a time and the number of different ways of climbing the stairs, S_n, is to be determined. If n is 1 then the solution is simple, $S_n = 1$. If $n = 2$ there are two ways, two single steps or one double step, i.e. $1 + 1$, or 2. For $n = 3$ there are three different ways; $1 + 2$, $2 + 1$ or $1 + 1 + 1$. This sequence can be generalized in the following way for $n > 2$. If a single step is taken initially then $n - 1$ stairs are left and there are S_{n-1} possibilities, if two steps are taken initially then $n - 2$ stairs are left corresponding to S_{n-2} possibilities. Thus the number of possibilities for n stairs is equal to the sum of S_{n-1} and S_{n-2}. That is;

$$S_n = S_{n-1} + S_{n-2}$$
(5.7)

which is equivalent to Eq. (5.3). This shows that the values of S_n follow the Fibonacci sequence with

$$S_n = F_{n+1}$$
(5.8)

and the values of S_n for small values of n as given above confirm this. The appearance of the Fibonacci numbers in this type of problem is indicative of its occurrence in a large number of statistical problems involving permutations and combinations.

An interesting property of Fibonacci numbers deals with Fibonacci squares as shown in Figs. 5.3 and 5.4. In Fig. 5.3 an odd number, n, of different Fibonacci rectangles are constructed (in this case 7) of dimensions F_i by F_{i+1} where i takes on values from 1 to $n + 1$. It is seen that these rectangles may be arranged to form a square with outer dimensions of F_{n+1} by F_{n+1}. The rectangles form a pattern which spirals outward much like the deflated golden rectangles of Fig. 3.5. Since the area of the square must be equal to the sum of the areas of the rectangles then Fig. 5.3 is a geometric proof of the relation

$$\sum_{i=2}^{n+1} F_i F_{i-1} = F_{n+1}^2 \qquad [n \ odd] \quad .$$
(5.9)

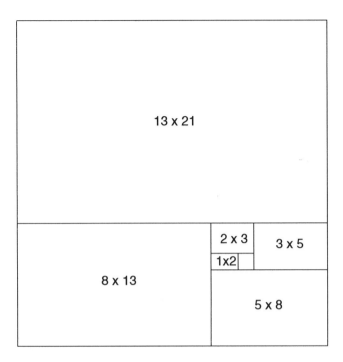

13 x 21

2 x 3

1x2

3 x 5

8 x 13

5 x 8

Fig. 5.3. A Fibonacci square comprised of an odd number of Fibonacci rectangles.

In Fig. 5.4 a similar Fibonacci square is constructed from an even number of Fibonacci rectangles. An area of dimensions 1×1 is left over indicating that the Fibonacci square has an area of 1 square unit larger than the sum of the areas of the rectangles. This is expressed as

$$\sum_{i=2}^{n+1} F_i F_{i-1} = F_{n+1}^2 - 1 \qquad [n \ even] \ . \qquad (5.10)$$

These are two of the fundamental mathematical relationships involving Fibonacci numbers which are presented in Appendix III.

An inspection of the Fibonacci numbers in Appendix II indicates that these numbers increase rapidly as a function of n. Although it is not readily apparent, there is a certain degree of periodicity in these numbers. Table 5.3 illustrates that the units digit of the Fibonacci numbers is cyclic with a periodicity of 60. That is, the units digit is the same as the units digit of F_{60} and also the same as the units digit of F_{120}. The same is true of F_1, F_{61} and F_{121}, etc. It appears as well that this

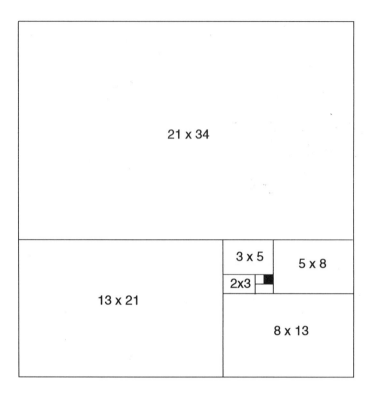

Fig 5.4. A Fibonacci square comprised of an even number of Fibonacci rectangles and an additional area of 1 × 1 unit.

periodicity continues indefinitely. It is found that there is also a periodicity in the ten's digits although the period is much longer. Similarly for the hundred's digits with an even longer period. It is speculated that this will extend to other digits of the Fibonacci numbers if sufficiently high values of n are investigated.

It is interesting to look at relationships between various Fibonacci numbers. Specifically the ratio of successive Fibonacci numbers is an interesting quantity. Table 5.4 gives some of these values. It is seen that as n increases then the ratio F_n/F_{n-1} approaches the golden ratio. Figure 5.5 shows that this ratio oscillates around the value of τ as a function of n and asymptotically approaches this value. This may be expressed as

Table 5.3. Periodicity of the units digit of Fibonacci numbers.

n	F_n	F_{n+60}	F_{n+120}
0	0	1548008755920	535835925499096664087 1840
1	1	2504730781961	86700073985079486580 51921
2	1	4052739537881	140283666653498915298923761
3	2	6557470319842	226983740520068639569 75682
4	3	10610209857723	367267407055057792558 99443
5	5	17167680177565	594251147575126432128 75125

Table 5.4. Ratios of successive Fibonacci numbers.

n	F_n	F_n/F_{n-1}
1	0	-
2	1	-
3	1	1.00000
4	2	2.00000
5	3	1.50000
6	5	1.66667
7	8	1.60000
8	13	1.62500
9	21	1.61539
10	34	1.61905
11	55	1.61768
12	89	1.61818
13	133	1.61798
14	233	1.61806
15	377	1.61803
16	610	1.61804

$$\lim_{n \to \infty} \frac{F_n}{F_{n-1}} = \tau \qquad (5.11)$$

and is a fundamental property of the Fibonacci sequence and the golden ratio. This relationship was first observed by Kepler, but a proof was not presented until more than a century later. The reasons for this property will not be discussed further here but will be clarified in Chapter 7. There are, however, further relationships between the Fibonacci numbers and the golden ratio; many of these are given in Appendix III and some will be shown later in this chapter to be of importance for calculating Fibonacci numbers.

The most obvious method of calculating a Fibonacci number F_n is to first calculate the Fibonacci numbers F_{n-1} and F_{n-2} and to add them together. This approach requires calculating all Fibonacci numbers of indices less than n before F_n may be calculated. The calculation of the Fibonacci numbers given in Table 5.1 by means of this method using a calculator or even by hand is straightforward.

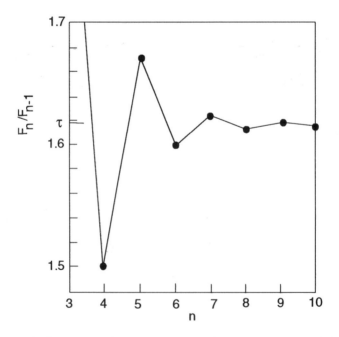

Fig 5.5. Ratio of successive Fibonacci numbers F_n/F_{n-1} as a function of n showing the convergence to the value of the golden ratio.

However, the calculation of the Fibonacci numbers given in Appendix II by this method becomes very tedious. The use of a computer greatly simplifies this task and the first 100 Fibonacci numbers can be generated in less than one second on an IBM Pentium. There are, however, some algorithms which can be utilized for the calculation of Fibonacci numbers with large n. A simple expression which is of use is

$$F_n = F_d F_{n+1-d} + F_{d-1} F_{n-d} \ .$$ (5.12)

If d is the integer equal to $n/2$ (or $(n+1)/2$ if n is odd) then only Fibonacci numbers up to (approximately) $n/2$ need to be calculated in order to obtain F_n. An even simpler and more elegant method relies upon the relationship between the Fibonacci numbers and the golden ratio. The relationship of importance is referred to as the Binet formula, after the French mathematician Jacques Phillipe Marie Binet [1786-1856] and may be written as

$$F_n = \frac{1}{\sqrt{5}} [\tau^n - (-\tau)^{-n}] \ .$$ (5.13)

Thus a value of F_n for any n may be calculated from the value of τ. Two factors, however, should be considered; (1) as τ is an irrational number and the Fibonacci number is an integer there will inevitably be round-off error introduced in this calculation and (2) if n is very large and a precise integer value of F_n is required then the value of τ which is used for the calculation must have a sufficient number of significant digits. From a practical standpoint it should be noted that if the second term on the right hand side of Eq. (5.13) is eliminated, the resulting expression;

$$F_n = \frac{\tau^n}{\sqrt{5}}$$ (5.14)

will be accurate to better than 1% of the value of F_n for $n > 4$. As this approximate value of F_n will oscillate about the true values of F_n as a function of n, an exact integer value of F_n may be obtained from

$$F_n = trunc[\frac{\tau^n}{\sqrt{5}} + \frac{1}{2}]$$ (5.15)

provided that the value of τ used for the calculation is sufficiently precise. The function *trunc* means the integer part of a number. These relationships (as well as many others given in Appendix III) clearly illustrate the relationship of the Fibonacci numbers to the golden ratio.

One final aspect of Fibonacci sequences which is of relevance here relates to the rabbit problem. A geometric sequence may be constructed on the basis of the rabbit breeding rules, as shown in Fig. 5.6, where a pair of adult rabbits is represented by a large rabbit icon and a pair of baby rabbits is represented by a small rabbit icon. Baby rabbit symbols are placed immediately to the right of their parents and the sequence progresses to the right. This arrangement of adult (A) and baby (b) rabbits may be written as

$$AbAAbAbAAbAAbAbAAbAbA \ \ldots \ . \tag{5.16}$$

Fig. 5.6. Geometric sequence of Fibonacci rabbits as a function of month. The large rabbit icon represents a pair of adult rabbits and the small icon a pair of baby rabbits. From Dunlap (1990).

This sequence of A's and b's is deterministic; that is, it may be extended indefinitely in a unique way because the rules for generating the next character in the sequence are well defined. In this sense it is distinct from a random sequence of A's and b's. However, it is also different from what is referred to as a periodic sequence. A simple example of a periodic sequence might be

$$Ab\!A\,bAbAbAbAbAbAbAb...\qquad\qquad(5.17)$$

where the pattern *Ab* is repeated indefinitely. There are several important characteristics of a periodic sequence; (1) a larger portion of the sequence may be generated by repeating a smaller portion of the sequence, in this case *Ab, AbAb, AbAbAb*, etc. may be repeated, and (2) these smaller portions of the sequence have the same ratio of *A*'s to *b*'s (two to one) as a larger portion of the sequence, including a sequence which is extended indefinitely. Neither of these characteristics applies to the Fibonacci sequence. One fundamental property of the Fibonacci sequence explains this behavior; the ratio of adult rabbits to baby rabbits in the limit of an infinite sequence is equal to the golden ratio, that is

$$\lim_{n\to\infty}\frac{A}{b}=\tau\quad.\qquad\qquad(5.18)$$

Since τ is an irrational number it cannot be represented by the ratio of two non-infinite rational numbers (or integers). This means that no finite portion of the Fibonacci rabbit sequence will have exactly the same ratio of adult to baby rabbits as the infinite sequence. Therefore, since repeating a finite portion of any sequence will produce a larger portion of a sequence with the same ratio of elements, a proper infinite Fibonacci sequence cannot be produced by repeating a finite sequence indefinitely. This type of sequence which is predictable but not periodic is referred to as aperiodic or, perhaps more properly, quasiperiodic.

Even a small portion of a Fibonacci sequence has some unique properties and it would be of interest to be able to distinguish a small portion of a Fibonacci sequence from a sequence of *A*'s and *b*'s which did not exhibit true quasiperiodicity. This problem has been discussed in detail by Penrose (1989). A portion of a Fibonacci sequence which begins at the beginning of the sequence shown in Eq. (5.16) is easily identified by generating a portion of the Fibonacci sequence of the same length and making a direct comparison. A portion of a sequence which does not begin at the beginning of the sequence in Eq. (5.16) is more difficult to analyze. Some obvious rules can be deduced from an inspection of Eq. (5.16); (1) there are no occurrences of the pattern *AAA* and (2) there are no occurrences of the pattern *bb*. Although these observations are based on the examination of a limited portion of a Fibonacci sequence, they are generally true as well. Thus, a simple inspection of the sequence to be tested for these patterns is informative. However, the nonexistence of these patterns in a sequence is no assurance of a proper Fibonacci sequence; Eq. (5.17) is an obvious example. A simple Fibonacci test based on the deflation of a sequence exists. The deflation rules are as follows;

(1) remove all isolated A's from the sequence
(2) replace all AA's by b's and
(3) replace all original b's by A's

These steps are repeated until either the null set results, in which case the original sequence is a valid Fibonacci sequence, or a forbidden pattern of elements occurs, in which case the original sequence was not a valid Fibonacci sequence. This test is based on the principle that the deflation of a Fibonacci sequence is another Fibonacci sequence. The application of this test to a valid Fibonacci sequence is shown in Table 5.5 and the application to a non-valid Fibonacci sequence is given in Table 5.6. An interesting aspect of the results illustrated in the tables is seen by observing the number of elements in the sequence as a function of the number of deflation operations. This is seen to follow a sequence of Fibonacci numbers. In cases where the initial number of elements in the sequence to be tested is not a Fibonacci number the reduction in length at each deflation will be slightly different from that given in the table until a Fibonacci number is encountered after which the numbers will follow the sequence of Fibonacci numbers. Thus the ratio of elements in the sequence before and after each deflation is given by F_n/F_{n-1}. As this ratio approaches the golden ratio in the limit of large n, this process may be

Table 5.5. Deflation of a valid Fibonacci sequence.

number	sequence	number of terms
1	AAbAAbAbAAbAA	13
2	bAbAAbAb	8
3	AAbAA	5
4	bAb	3
5	AA	2
6	b	1
7	A	1
8	{nul}	0

referred to, as in the case of the rescaling of the golden triangles or golden rectangles, as a deflation by a factor of τ. More will be said about this in Chapter 10.

Table 5.6. Deflation of a non-valid Fibonacci sequence.

number	sequence	number of terms
1	*AAbAAbAAbAbAA*	13
2	*bAbAbAAb*	8
3	*AAAbA*	5

CHAPTER 6

LUCAS NUMBERS AND GENERALIZED FIBONACCI NUMBERS

The discussion in the previous chapter demonstrated the relationship of the golden ratio to the additive sequence known as the Fibonacci sequence. This is generated by the additive recursion relation (Eq. (2.1)) using the seed values $F_0 = 0$ and $F_1 = 1$. Numerous other additive sequences can be formed by using different seed values. One, given by Eq. (2.14), has already been discussed. In the present chapter some additional possibilities for integer seed values will be considered. Using integers which are any consecutive terms in a Fibonacci sequence as seed values (e.g. 1,2; 2,3; 3,5 etc.) will merely yield a Fibonacci sequence with the values of the indices shifted. Using seed values which are multiples of Fibonacci numbers (e.g. 0,2; 0,3; 2,2; 2,4; 6,10 etc.) will merely produce a Fibonacci sequence with all terms multiplied by a constant factor. Most of what has been discussed in the previous chapter will apply to such sequences, particularly the fact that the ratio F_n/F_{n-1} will approach τ in the limit of large n. Certain choices of seed values will, however, yield additive sequences which are distinctly different from the Fibonacci sequence. One such possibility was studied extensively in the late nineteenth century by the French mathematician Edouard Lucas who published the results of these investigations in 1877. He considered the next smallest seed values $L_0 = 2$ and $L_1 = 1$. These values will generate the additive sequence

$$2, \ 1, \ 3, \ 4, \ 7, \ 11, \ 18, \ 29, \ 47, \ 76, \ 123, \ \dots \ . \tag{6.1}$$

Analogous to the discussion concerning Eq. (2.3), the Lucas sequence can be extended to negative indices as well to yield

$$\dots \ 7, \ -4, \ 3, \ -1, \ 2, \ 1, \ 3, \ 4, \ 7, \ \dots \ . \tag{6.2}$$

In general, a comparison of Eqs. (2.3) and (6.2) allows for the determination of general expressions for Fibonacci and Lucas numbers with negative indices;

$$F_{-n} = (-1)^{n+1} F_n \qquad (6.3)$$

and

$$L_{-n} = (-1)^n L_n \ . \qquad (6.4)$$

Table 6.1 gives values of some Lucas numbers and the ratio L_n/L_{n-1}. It is seen from the values in the table that the ratio of Lucas numbers approaches the golden ratio as n becomes large.

The so-called generalized Fibonacci numbers, G_n, are produced from the recursion relation for an additive sequence with arbitrary values of the seeds. In general terms, for $G_0 = p$ and $G_1 = q$, the sequence

$$p, \ q, \ p+q, \ p+2q, \ 2p+3q, \ 3p+5q, \ ... \qquad (6.5)$$

Table 6.1. Ratios of successive Lucas numbers.

n	L_n	L_n/L_{n-1}
0	2	-
1	1	0.50000
2	3	3.00000
3	4	1.33333
4	7	1.75000
5	11	1.57143
6	18	1.63636
7	29	1.61111
8	47	1.62069
9	76	1.61702
10	123	1.61842
11	199	1.61789
12	322	1.61809
13	521	1.61801
14	843	1.61804
15	1364	1.61803

is produced. It is easy to see that the Fibonacci numbers are the coefficients of the p's and q's in the terms of this sequence. That is,

$$G_{n+2} = F_n p + F_{n+1} q \quad . \tag{6.6}$$

It is of interest to consider an example of the generalized Fibonacci sequence. For seed values $G_0 = 7$ and $G_1 = 3$ the values of the terms of the additive sequence are given in Table 6.2. Again it is seen that the ratio G_n/G_{n-1} approaches the golden ratio for large n. It is, in fact, true that the ratio of successive terms in any additive sequence approaches the golden ratio for large n, and the mathematical reasons for this will be demonstrated in the next chapter. If non-integer or even non-rational values are permitted as seed values then this behavior still occurs. An inspection of the additive sequence in Eq. (2.14) (and a comparison with Eq. (2.13)) indicates that for seed values $G_0 = 1$ and $G_1 = \tau$, the ratio of terms in the sequence is τ for all values of n.

Table 6.2. Ratios of successive numbers in the additive sequence generated by seed values $G_0 = 7$ and $G_1 = 3$.

n	G_n	G_n/G_{n-1}
0	7	-
1	3	0.42857
2	10	3.33333
3	13	1.30000
4	23	1.76923
5	36	1.56522
6	59	1.63889
7	95	1.61017
8	154	1.62105
9	249	1.61688
10	403	1.61847
11	652	1.61787
12	1055	1.61810
13	1707	1.61801
14	2762	1.61804
15	4469	1.61803

The calculation of Lucas numbers follows very closely along the lines of the discussion in the previous chapter concerning the calculation of Fibonacci numbers. The simple, straightforward method of calculating L_n is to calculate all previous Lucas numbers and use the appropriate recursion relation. The relationship of Lucas numbers to the golden ratio may be used as a means of calculating L_n. An expression which demonstrates the relationship between the Lucas numbers and the golden ratio is

$$L_n = \tau^n - (-\tau)^{-n} \ . \tag{6.7}$$

A more practical expression for actually calculating values of L_n is based on the fact that the second term on the right hand side of Eq. (6.7) is negligible for n greater than about 4;

$$L_n = trunc[\tau^n + \frac{1}{2}] \ . \tag{6.8}$$

Table 6.3. Ratios of Fibonacci and Lucas numbers.

n	F_n	L_n	$\sqrt{5}\,F_n/L_n$
0	0	2	0.00000
1	1	1	2.23607
2	1	3	0.74536
3	2	4	1.11803
4	3	7	0.95831
5	5	11	1.01639
6	8	18	0.99381
7	13	29	1.00238
8	21	47	0.99909
9	34	76	1.00035
10	55	123	0.99987
11	89	199	1.00005
12	144	322	0.99998
13	233	521	1.00001
14	377	843	1.00000
15	610	1364	1.00000

The above relations can be compared with those for Fibonacci numbers discussed in the previous chapter. A simple relationship between Fibonacci and Lucas numbers can be obtained of the form

$$L_n = \sqrt{5}F_n \; . \tag{6.9}$$

Table 6.3 illustrates the validity of this relationship for values of n greater than 4 or 5. As well, a relationship which is useful in calculating large Fibonacci numbers is

$$F_{2n} = F_n L_n \; . \tag{6.10}$$

Another interesting property of Fibonacci and Lucas numbers deals with divisibility. A simple inspection of the first few Fibonacci numbers given in Appendix II, reveals the following general trends;

(1) $F_3 = 2$ and this evenly divides all other even Fibonacci numbers. An inspection of the table shows that these are $F_6 = 8$, $F_9 = 34$, $F_{12} = 144$, $F_{15} = 610$, ... etc.
(2) $F_4 = 3$ and this divides the Fibonacci numbers $F_8 = 21$, $F_{12} = 144$, $F_{16} = 987$, $F_{20} = 6765$, ... etc.

These trends are easily extended and may be expressed as the following simple theorem for $n > 1$:

F_n **divides** F_m **if and only if** $m = kn$ $(k = 1,2,3...)$.

An inspection of the Lucas numbers in Appendix II does not reveal such obvious relationships between the L_n. However, it can be shown (Carlitz 1964) that an analogous theorem for the Lucas numbers may be expressed (for $n > 1$) as

L_n **divides** L_m **if and only if** $m = (2k-1)n$ $(k = 1,2,3...)$.

Examples of the validity of this theorem are;

(1) $F_2 = 3$ divides $F_6 = 18$, $F_{10} = 123$, ... etc. and
(2) $F_3 = 4$ divides $F_9 = 76$, $F_{15} = 1364$, ... etc.

It can also be shown that a similar relationship exists between the Fibonacci and Lucas numbers and may be expressed (for $n > 1$) as

L_n **divides** F_m **if and only if** $m = 2kn$ $(k = 1,2,3...)$.

The calculation of Fibonacci or Lucas numbers in different moduli is a topic which is closely related to divisibility as this deals with the remainder of the division process. The values of $F_n(\text{mod } m)$ for $2 \leq m \leq 9$ and $0 \leq n \leq 27$ are given in Table 6.4. It is readily seen from this example that $F_n(\text{mod } m)$ is periodic. In fact, this behavior has already been seen in the previous chapter where the units digit was shown to have a periodicity of 60. It should be noted that the units digit of F_n is merely $F_n(\text{mod } 10)$. The length of the repeat cycle for $F_n(\text{mod } m)$ for some values of m is given in Table 6.5. Similar periodic behavior is observed for Lucas numbers as given in Table 6.6. As shown in Table 6.5, the periodicity of Lucas numbers is sometimes, but not always, the same as a function of m as is found for the Fibonacci numbers. This behavior may be extended to higher values of m, but, as suggested in Chapter 5, the repeat cycle becomes longer as m increases. These concepts may also be extended to include periodic behavior of the digits of the generalized Fibonacci numbers.

The problem of divisibility of a set of numbers also raises the question of prime factors. It may seem that the theorem concerning the divisibility of Fibonacci numbers given above may imply that F_n is prime if n is prime. It is, however, readily apparent that $F_4 = 3$ is prime while $n = 4$ is not prime. It can be proved that F_4 is the only Fibonacci number which is a prime for a value of n which is not prime (Vajda 1989). Thus for F_n ($n \neq 4$) to be a prime it is necessary for n to be a prime. However, this condition is not sufficient for F_n to be prime. This is seen from the value of $F_{19} = 4181 = 37 \times 113$. Thus the question of which Fibonacci numbers are prime is far from straightforward, although only values of F_n for which n is prime need be considered as possible candidates. An inspection of the values of Lucas numbers in Appendix II indicates that the question of prime factors of L_n is even more difficult. Table 6.7 gives the prime factors of Fibonacci and Lucas numbers for values of $n \leq 30$. It is not known whether the infinite sequences of Fibonacci and Lucas numbers contain an infinite number of primes. It has, however, been claimed (see Schroeder 1984) that the infinite generalized Fibonacci sequence produced with seed values

$$G_0 = 1786772701928802632268715130455793$$
$$G_1 = 1059683225053915111058165141686995$$

(6.11)

contains no prime at all. The proof of this hypothesis is not known.

The concept of an additive sequence can be extended to include other recursion relations than that given in Eq. (2.1). One natural extension is to consider the sequence of numbers where each term is the sum of the previous three

Table 6.4. Values of Fibonacci numbers in modulus m, F_n(mod m). The repeat cycle for the periodicity is shown in bold face.

n	F_n	$m=2$	$m=3$	$m=4$	$m=5$	$m=6$	$m=7$	$m=8$	$m=9$
					F_n(mod m)				
0	0	**0**	**0**	**0**	**0**	**0**	**0**	**0**	**0**
1	1	**1**	**1**	**1**	**1**	**1**	**1**	**1**	**1**
2	1	**1**	**1**	**1**	**1**	**1**	**1**	**1**	**1**
3	2	0	**2**	**2**	**2**	**2**	**2**	**2**	**2**
4	3	1	**0**	**3**	**3**	**3**	**3**	**3**	**3**
5	5	1	**2**	**1**	0	**5**	**5**	**5**	**5**
6	8	0	**2**	0	3	**2**	**1**	**0**	**8**
7	13	1	**1**	1	3	**1**	**6**	**5**	**4**
8	21	1	**0**	1	**1**	**3**	**0**	**5**	**3**
9	34	0	1	2	4	**4**	**6**	**1**	**7**
10	55	1	1	3	**0**	**1**	**6**	**7**	**1**
11	89	1	2	1	4	**5**	**5**	**1**	**8**
12	144	0	0	0	4	**0**	**4**	**0**	**0**
13	233	1	2	1	3	**5**	**2**	**1**	**7**
14	377	1	2	1	2	**5**	**6**	**1**	**8**
15	610	0	1	2	0	**4**	**1**	**2**	**7**
16	987	1	0	3	2	**3**	**0**	**3**	**6**
17	1597	1	1	1	2	**1**	**1**	**5**	**4**
18	2584	0	1	0	4	**4**	**1**	**0**	**1**
19	4181	1	2	1	1	**5**	**2**	**5**	**5**
20	6765	1	0	1	0	**3**	**3**	**5**	**6**
21	10946	0	2	2	1	**2**	**5**	**1**	**2**
22	17711	1	2	3	1	**5**	**1**	**7**	**8**
23	28657	1	1	1	2	**1**	**6**	**1**	**1**
24	46368	0	0	0	3	**0**	**0**	**0**	**0**
25	75025	1	1	1	0	**1**	**6**	**1**	**1**
26	121393	1	1	1	3	**1**	**6**	**1**	**1**
27	196418	0	2	2	3	**2**	**5**	**2**	**2**

terms. This is expressed as

$$T_n = T_{n-1} + T_{n-2} + T_{n-3} \tag{6.12}$$

and numbers which obey this relation are referred to as Tribonacci numbers.
Three seed values are required to generate such a sequence, and, in a manner
analogous to the various problems involving Fibonacci numbers, a natural choice
for seed values to investigate the properties of Tribonacci numbers might be $T_0 = 0$,
$T_1 = 0$ and $T_2 = 1$. These seed values will yield the sequence given in Table 6.8.

Along the lines of the previous discussions it is of interest to consider the ratio
of successive terms in this sequence. This is given in the table. This ratio appears
to converge to a well defined value, although not the golden ratio. For large n this
value is found to be $T_n/T_{n-1} =$ 1.8392867552 1416113255 1852564653 2866004241
787460975 ... . The mathematical reason that the ratio of Fibonacci numbers
approaches the golden ratio will be discussed in detail in Chapter 7. The reason
the ratio of Tribonacci numbers approaches the value given above is beyond the
scope of this book. It is possible, however, to gain some insight into the origins of
this number. It can be recalled that the value of the golden ratio satisfies the

Table 6.5. Length of the repeat cycle for $F_n(\text{mod } m)$ and $L_n(\text{mod } m)$.

	periodicity	
m	F_n	L_n
1	1	1
2	3	3
3	8	8
4	6	6
5	20	4
6	24	24
7	16	16
8	12	12
9	24	24
10	60	12
11	10	10
12	24	24

Table 6.6. Values of Lucas numbers in modulus m, $L_n(\bmod m)$. The repeat cycle for the periodicity is shown in bold type.

n	L_n	$L_n(\bmod m)$							
		$m=2$	$m=3$	$m=4$	$m=5$	$m=6$	$m=7$	$m=8$	$m=9$
0	2	**0**	**2**	**2**	2	**2**	2	2	2
1	1	**1**	**1**	**1**	1	**1**	1	1	1
2	3	**1**	**0**	3	3	3	3	3	3
3	4	**0**	**1**	0	4	4	4	4	4
4	7	1	**1**	3	2	1	0	7	7
5	11	1	**2**	3	1	5	4	3	2
6	18	0	**0**	2	3	**0**	4	2	**0**
7	29	1	**2**	1	4	5	1	5	2
8	47	1	2	3	2	5	5	7	2
9	76	0	1	0	1	4	6	4	4
10	123	1	0	3	3	3	4	3	6
11	199	1	1	3	4	**1**	3	7	**1**
12	322	0	1	2	2	**4**	0	2	7
13	521	1	2	1	1	5	3	1	8
14	843	1	0	3	3	3	3	3	6
15	1364	0	2	0	4	**2**	6	4	5
16	2207	1	2	3	2	5	2	7	2
17	3571	1	1	3	1	**1**	1	3	7
18	5778	0	0	2	3	**0**	3	2	**0**
19	9349	1	1	1	4	**1**	4	5	7
20	15127	1	1	3	2	**1**	0	7	7
21	24476	0	2	0	1	**2**	4	4	5
22	39603	1	0	3	3	**3**	4	3	**3**
23	64079	1	2	3	4	5	1	7	8
24	103682	0	2	2	2	2	5	2	2
25	167761	1	1	1	1	1	6	1	1
26	271443	1	0	3	3	3	4	3	3
27	439204	0	1	0	4	4	3	4	4

Table 6.7. Prime factors of Fibonacci and Lucas numbers. Primes are given in bold face.

n	F_n	L_n
0	0	2
1	1	1
2	1	3
3	2	2^2
4	3	7
5	5	11
6	2^3	2×3^2
7	13	29
8	3×7	47
9	2×17	$2^2 \times 19$
10	5×11	3×41
11	89	199
12	$2^2 \times 3^2$	$2 \times 7 \times 23$
13	233	521
14	13×29	3×281
15	$2 \times 5 \times 61$	$2^2 \times 11 \times 31$
16	$3 \times 7 \times 47$	2207
17	1597	3571
18	$2^3 \times 17 \times 19$	$2 \times 3^3 \times 107$
19	37×113	9349
20	$3 \times 5 \times 11 \times 41$	7×2161
21	$2 \times 13 \times 421$	$2^2 \times 29 \times 211$
22	89×199	$3 \times 43 \times 307$
23	28657	139×461
24	$2^5 \times 3^2 \times 7 \times 23$	$2 \times 47 \times 1103$
25	$5^2 \times 3001$	$11 \times 101 \times 151$
26	233×521	3×90481
27	$2 \times 17 \times 53 \times 109$	$2^2 \times 19 \times 5779$
28	$3 \times 13 \times 29 \times 281$	$7^2 \times 14503$
29	514229	59×19489
30	$2^3 \times 5 \times 11 \times 31 \times 61$	$2 \times 3^2 \times 41 \times 2521$

Table 6.8. The Tribonacci numbers and their ratios.

n	T_n	T_n/T_{n-1}
0	0	-
1	0	-
2	1	-
3	1	1.00000
4	2	2.00000
5	4	2.00000
6	7	1.75000
7	13	1.85714
8	24	1.84615
9	44	1.83333
10	81	1.80491
11	149	1.83951
12	274	1.83893
13	504	1.83942
14	927	1.83929
15	1705	1.83927

Fibonacci quadratic equation given by Eq. (2.9). The analogous expression which is of relevance for the Tribonacci sequence is the Fibonacci (or Tribonacci) cubic equation given by (Dunlap 1996)

$$x^3 - x^2 - x - 1 = 0 \ . \tag{6.13}$$

It can be shown that this equation has two imaginary roots and one real root (see Abramowitz and Stegun 1964). The real root is given by

$$x = \frac{1}{3}\left[\left(19 + 3\sqrt{33}\right)^{1/3} + \left(19 - 3\sqrt{33}\right)^{1/3} + 1\right] \tag{6.14}$$

which has the value 1.839287 ... . This generalization of the additive sequence can be extended with predictable results.

CHAPTER 7

CONTINUED FRACTIONS AND RATIONAL APPROXIMANTS

Any irrational number, I, can be expressed in terms of an infinite number of integers, a_0, a_1, a_2 ... in the form

$$I = a_0 + \cfrac{1}{a_1 + \cfrac{1}{a_2 + \cfrac{1}{a_3 + \cfrac{1}{a_4 + \cfrac{1}{a_5 + \dots}}}}} \quad . \tag{7.1}$$

This kind of expression is called a continued fraction (Schroeder 1984) and can be written in more compact form as

$$I = [a_0, a_1, a_2, a_3, a_4, a_5, \dots] \quad . \tag{7.2}$$

The golden ratio may be expressed in this form if the proper values of all of the a_i's are determined. This is relatively straightforward. Consider the quadratic equation given by Eq. (2.23);

$$x^2 + x - 1 = 0 \quad . \tag{7.3}$$

The positive solution of this equation is $x = 1/\tau$. The quadratic equation may be rewritten as

$$x(x+1) = 1 \tag{7.4}$$

or

$$x = \frac{1}{1+x} \quad . \tag{7.5}$$

The x on the right hand side of the equation can be replaced by $1/(1+x)$ to give

$$x = \cfrac{1}{1+\cfrac{1}{1+x}} \ . \tag{7.6}$$

This substitution process may be continued indefinitely to give

$$x = \cfrac{1}{1+\cfrac{1}{1+\cfrac{1}{1+\cfrac{1}{1+...}}}} \ . \tag{7.7}$$

Since the solution to Eq. (7.3) is $1/\tau$, Eq. (7.7) becomes

$$\frac{1}{\tau} = \cfrac{1}{1+\cfrac{1}{1+\cfrac{1}{1+\cfrac{1}{1+...}}}} \ . \tag{7.8}$$

The inverse of the golden ratio is related to τ by the simple relation

$$\tau = 1 + \frac{1}{\tau} \ , \tag{7.9}$$

then Eq. (7.8) gives the golden ratio as

$$\tau = 1 + \cfrac{1}{1+\cfrac{1}{1+\cfrac{1}{1+\cfrac{1}{1+...}}}} \ . \tag{7.10}$$

A simple comparison of Eqs. (7.1) and (7.10) shows that

$$\tau = [a_0, a_1, a_2, a_3, a_4, a_5, ...] = [1,1,1,1,1,1,...] \ . \tag{7.11}$$

In an actual calculation, an infinite number of terms in the continued fraction form for τ cannot be included. Therefore, an actual calculation would terminate after some finite number of terms, n, and, the continued fraction would be expressed as

$[a_0, a_1, a_2,...,a_n]$. This representation of an irrational number is referred to as the nth rational approximant. The rational approximants of the golden ratio are merely $[1,1,1,1,...,1]$ and are given as a function of n (where n is the number of 1's minus 1) in Table 7.1. A comparison of the values in this table with those of Table 5.4 shows that the nth rational approximate of the golden ratio, τ_n, is merely

$$\tau_n = \frac{F_{n+1}}{F_n} . \qquad (7.12)$$

It is obvious now why the ratio of successive Fibonacci numbers approaches the golden ratio in the limit of large n.

This same approach may be taken with the sequence of Lucas numbers, and the rational approximants may be expressed as in Table 7.2. A consideration of the rational representation in Table 7.2 in the context of the continued fraction representation shows that

$$\frac{1}{2} = \frac{1}{2}$$

$$\frac{3}{1} = 1 + \frac{1}{1/2}$$

$$\frac{4}{3} = 1 + \cfrac{1}{1 + \cfrac{1}{1/2}}$$

$$\frac{7}{4} = 1 + \cfrac{1}{1 + \cfrac{1}{1 + \cfrac{1}{1/2}}} . \qquad (7.13)$$

Thus it is easy to see that the nth rational approximant based on the Lucas numbers is given by $[a_0, a_1, a_2,...,a_n] = [1,1,...,1/2]$. Clearly as n becomes large the importance of the fact that the final term in the rational approximant is 1/2 rather than 1 as it is for the Fibonacci sequence becomes vanishingly small, and it is apparent that the ratio of Lucas numbers will also approach the golden ratio in the

Table 7.1. Rational approximants of the golden ratio.

n	rational representation	decimal value
0	1	1.00000
1	2/1	2.00000
2	3/2	1.50000
3	5/3	1.66666
4	8/5	1.60000
5	13/8	1.62500
6	21/13	1.61539
7	34/21	1.61905
8	55/34	1.61768
9	89/55	1.61818
10	144/89	1.61798

Table 7.2. Rational approximants for the golden ratio based on the Lucas numbers.

n	rational representation	decimal value
0	1/2	0.50000
1	3/1	3.00000
2	4/3	1.33333
3	7/4	1.75000
4	11/7	1.57143
5	18/11	1.63636
6	29/18	1.61111
7	47/29	1.62069
8	76/47	1.61702
9	123/76	1.61842
10	199/123	1.61789

limit of large n. The final additive sequence which was discussed in detail in Chapter 6, i.e. $G_0 = 7$ and $G_1 = 3$, may also be considered. Table 7.3 shows the rational approximants based on this sequence. In terms of the continued fraction representation it is seen that

$$\frac{3}{7} = \frac{3}{7}$$

$$\frac{10}{3} = 1 + \frac{1}{3/7}$$

$$\frac{13}{10} = 1 + \cfrac{1}{1 + \cfrac{1}{3/7}}$$

$$\frac{23}{13} = 1 + \cfrac{1}{1 + \cfrac{1}{1 + \cfrac{1}{3/7}}} \quad . \tag{7.14}$$

Thus the nth rational approximant based on this additive sequence is given by $[a_0, a_1, a_2,...,a_n] = [1,1,...,3/7]$. Again as n becomes large the importance of a_n becomes small and it is clear why the ratio of terms in this additive sequence also approaches the golden ratio. On the basis of the evidence presented above it is possible to draw some general conclusions concerning the ratios of the terms in an additive sequence. For seed values $G_0 = p$ and $G_1 = q$, the nth rational approximant is expressed as

$$[a_0, a_1, a_2, ..., a_n] = [1,1,1,..., \frac{q}{p}] \quad . \tag{7.15}$$

This relationship may be shown to be valid in a somewhat more rigorous manner by considering the ratio of terms in a generalized Fibonacci sequence, G_{n+1}/G_n. From the definition of the additive sequence it is known that $G_{n+1} = G_n + G_{n-1}$, so the ratio may be expressed as

$$\frac{G_{n+1}}{G_n} = \frac{G_n + G_{n-1}}{G_n} = 1 + \frac{G_{n-1}}{G_n} \tag{7.16}$$

Table 7.3. Rational approximants for the golden ratio based on the additive sequence with $G_0 = 7$ and $G_1 = 3$.

n	rational representation	decimal value
0	3/7	0.42857
1	10/3	3.33333
2	13/10	1.30000
3	23/13	1.76923
4	36/23	1.56522
5	59/36	1.63889
6	95/59	1.61017
7	154/95	1.62105
8	249/154	1.61688
9	403/249	1.61847
10	652/403	1.61787

or

$$\frac{G_{n+1}}{G_n} = 1 + \frac{1}{G_n / G_{n-1}} \quad . \tag{7.17}$$

Rewriting G_n/G_{n-1} as

$$\frac{G_n}{G_{n-1}} = 1 + \frac{1}{G_{n-1} / G_{n-2}} \tag{7.18}$$

yields

$$\frac{G_{n+1}}{G_n} = 1 + \frac{1}{1 + \dfrac{1}{G_{n-1} / G_{n-2}}} \quad . \tag{7.19}$$

Continuing this substitution process leads to

$$\frac{G_{n+1}}{G_n} = 1 + \cfrac{1}{1 + \cfrac{1}{1 + \cfrac{1}{1 + \cfrac{1}{...G_1/G_0}}}} \tag{7.20}$$

or merely

$$\frac{G_{n+1}}{G_n} = [1,1,1,\ldots,G_1/G_0] \quad . \tag{7.21}$$

In the case of the Fibonacci sequence, $G_0 = 0$ and the term involving G_0 will vanish leaving the $G_2/G_1 = 1/1$ term as the lowest order remaining term.

The accuracy with which an irrational number may be approximated by its rational approximate may be judged by comparing the actual value of the irrational number, I, to G_{n+1}/G_n. Certainly as n increases the difference $\Delta(n) = |I - G_{n+1}/G_n|$ will decrease. However, for a given value of n it can be shown that among all irrational numbers, $\Delta(n)$ is the largest for the golden ratio. This means that if we were to assign a measure of the degree of irrationality of an irrational number on the basis of the quantity $\Delta(n)$, we would conclude that the golden ratio was the most irrational of all irrational numbers.

A physical system which demonstrates the principles of continued fractions and rational approximants is the infinite resistor network shown in Fig. 7.1. The total resistance of the network, R, is given by

$$R = [R_1, R_2, R_3, \ldots] \quad . \tag{7.22}$$

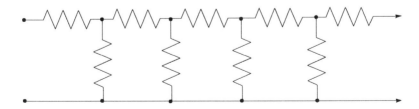

Fig. 7.1. An infinite resistor network.

In the case where all resistors are 1 *ohm* then the resistance is given by $R = [1,1,1,...]$ and from Eq. (7.10) a resistance of τ *ohms* is obtained. Further discussion of this problem can be found in March (1993) and Srinivasan (1992).

The above results have justified the observations which were presented in Chapters 5 and 6 concerning the ratios of successive terms in an additive sequence and have demonstrated the relevance of the golden ratio to the properties of a generalized Fibonacci sequence. The concept of rational approximants has been introduced and this will be seen to play an important role in applications of the golden ratio discussed in later chapters.

CHAPTER 8

GENERALIZED FIBONACCI REPRESENTATION THEOREMS

The representation theorems discussed in this chapter are based on the fact that a generalized Fibonacci sequence forms a complete set. A sequence of integers is said to be complete if any positive integer can be expressed as a sum of a finite number of the terms in the sequence and each term is used, at most, once. Before dealing with generalized Fibonacci sequences in this respect, a commonly known complete set will provide an informative example of representation theorems.

The binary number system is based on the fact that the powers of 2, (i.e. 2^0, 2^1, 2^2, 2^3, ...), form a complete set. Any integer, M, may be expressed as a sum of a finite number of terms involving the powers of 2 and appropriate coefficients. This is expressed as

$$M = \sum_{i=0}^{N} \sigma_i 2^i \qquad (8.1)$$

where the coefficients, σ_i, are 0 or 1. The value of N is chosen such that

$$2^N \le M < 2^{N+1} . \qquad (8.2)$$

As a simple example of the binary system the number 12 may be represented as 1100 since

$$12 = 1 \times 2^3 + 1 \times 2^2 + 0 \times 2^1 + 0 \times 2^0 . \qquad (8.3)$$

N is chosen to be 3 as $2^3 \le 12 < 2^4$ (i.e. $8 \le 12 < 16$). The set of powers of 2 with any one term removed is not complete as it does not allow for the binary representation of all integers.

The set of Fibonacci numbers (F_2, F_3, F_4, ...) forms a complete set. The inclusion of $F_0 = 0$ is unnecessary and the inclusion of $F_1 = 1$ is redundant. The set of Fibonacci numbers (F_2, F_3, F_4, ...) with any one term removed is not complete, although the inclusion of F_1 allows for the arbitrary removal of any single other term without loss of completeness. The set of Fibonacci numbers (F_2, F_3, F_4, ...) will be considered as a complete set for the demonstration of representation theorems in the present chapter.

Any positive integer may be represented as a finite sum of Fibonacci numbers with appropriate coefficients $\sigma_i = (0,1)$ as

$$M = \sum_{i=2}^{N} \sigma_i F_i \quad . \tag{8.4}$$

The choice of N is not so well defined as in the case of the binary system. Choosing N such that

$$I \le F_N \tag{8.5}$$

is a sufficient condition although a smaller value of N may, in some cases, be suitable as will be shown below. As an example of Fibonacci representation the number 12 is considered. It is easy to see that 12 may be written as

$$12 = 1 \times F_6 + 0 \times F_5 + 1 \times F_4 + 0 \times F_3 + 1 \times F_2 \quad . \tag{8.6}$$

Thus in Fibonacci representation the number 12 is written as 10101. Table 8.1 shows possible Fibonacci representations for the first 20 positive integers. The results illustrated here are distinct from those which are obtained from the powers of 2 representation as Table 8.1 clearly shows that certain integers (in fact most of them) can be represented in more than one way by a sum of Fibonacci numbers. A few integers have only one representation and these are seen to follow a particular pattern. In fact it can be shown that a single Fibonacci representation of M exists if and only if

$$M = F_N - 1 \qquad [n = 1,2,3,...] \tag{8.7}$$

If Fibonacci numbers are to form the basis of a useful system of integer representation it would be desirable to have a unique representation for each integer. An inspection of the data in Table 8.1 shows that the various representations for the same integer may be classified on the basis of the number of 1's which are present in the Fibonacci representation. The representation which contains the smallest number of 1's is called the minimal or canonical representation of the integer and the table shows that this is, in each case, unique, at least for integers up to 20. It can be shown that this is a general property of

Table 8.1. Fibonacci representations for the integers 1 to 20.

integer	Fibonacci representation						
1	1						
2	10						
3	100	or	11				
4	101						
5	1000	or	110				
6	1001	or	111				
7	1010						
8	10000	or	1100	or	1011		
9	10001	or	1101				
10	10010	or	1110				
11	10100	or	10011	or	1111		
12	10101						
13	100000	or	11000	or	10110		
14	100001	or	11001	or	10111		
15	100010	or	11010				
16	100100	or	100011	or	11100	or	11011
17	100101	or	11101				
18	101000	or	100110	or	11110		
19	101001	or	100111	or	11111		
20	101010						

Fibonacci representations. Thus the condition that the canonical representation should be chosen over other Fibonacci representations will yield a unique representation for each positive integer. An interesting property of the canonical representation known as Zeckendorf's theorem (see Vajda 1989) is helpful in immediately identifying the minimal representation for an integer. This states that

$$\sigma_i \sigma_{i+1} = 0 \qquad [i = 1,2,3,\ldots,N-1] \quad . \tag{8.8}$$

This may be stated in the alternate way; no two consecutive digits in the canonical representation are non-zero. This condition is both sufficient and necessary for identifying the canonical representation. Thus the application of Zeckendorf's theorem allows for the determination of a unique Fibonacci representation for each positive integer as given in Table 8.2.

Extending the above representation theorems to the Lucas numbers it can be shown that $(L_0, L_1, L_2, ...)$ forms a complete set. L_1 is required because 1 is a required element in any complete set, L_0 is required because 2 is a necessary element in any complete set (unless there are two 1's). Because these terms are a requirement for completeness then the Lucas sequence with any one arbitrary element removed is not complete. For the complete Lucas sequence each integer can be represented as

$$M = \sum_{i=0}^{N} \sigma_i L_i \qquad (8.9)$$

where

$$M \le L_n \qquad (8.10)$$

is a sufficient but not stringent condition on the determination of N. The Lucas representations of the integers from 1 to 20 are given in Table 8.3. As with the Fibonacci representations, these are not unique. In fact they are not entirely analogous to the Fibonacci representations as only 1 (represented as 10) and 2 (represented as 1) are unique. The consideration of the minimal representation does not resolve this ambiguity as 5 and 16 (as well as larger integers not given in the table) have more than one minimal representation. The application of Zeckendorf's theorem does not resolve this problem either as the two minimal representations of 5 and 16 both satisfy Eq. (8.8). Some additional constraint must be imposed on Lucas representations in order to insure uniqueness. Equation (8.8) is sufficient for this purpose if it is also required that

$$\sigma_0 \sigma_2 = 0 \quad . \qquad (8.11)$$

That is, both L_0 and L_2 are not used in the same representation. Thus for 5, the correct Lucas representation is 1010 rather than 101 and for 16 it is 101010 rather than 100101. The canonical Lucas representations for the first 40 positive integers are given in Table 8.2.

Table 8.2. Decimal, binary, canonical Fibonacci and canonical Lucas representations of the first 40 positive integers.

decimal	binary	Fibonacci	Lucas
1	1	1	10
2	10	10	1
3	11	100	100
4	100	101	1000
5	101	1000	1010
6	110	1001	1001
7	111	1010	10000
8	1000	10000	10010
9	1001	10001	10001
10	1010	10010	10100
11	1011	10100	100000
12	1100	10101	100010
13	1101	100000	100001
14	1110	100001	100100
15	1111	100010	101000
16	10000	100100	101010
17	10001	100101	101001
18	10010	101000	1000000
19	10011	101001	1000010
20	10100	101010	1000001
21	10101	1000000	1000100
22	10110	1000001	1001000
23	10111	1000010	1001010
24	11000	1000100	1001001
25	11001	1000101	1010000
26	11010	1001000	1010010
27	11011	1001001	1010001
28	11100	1001010	1010100
29	11101	1010000	10000000
30	11110	1010001	10000010
31	11111	1010010	10000001
32	100000	1010100	10000100

Table 8.2. continued

33	100001	1010101	10001000
34	100010	10000000	10001010
35	100011	10000001	10001001
36	100100	10000010	10010000
37	100101	10000100	10010010
38	100110	10000101	10010001
39	100111	10001000	10010100
40	101000	10001001	10100000

Table 8.3. Lucas representations of the first 20 positive integers.

integer	Lucas representation						
1	10						
2	1						
3	100	or	11				
4	1000	or	110				
5	1010	or	101				
6	1001	or	111				
7	10000	or	1100	or	1011		
8	¹0010	or	1110				
9	10001	or	1101				
10	10100	or	10011	or	1111		
11	100000	or	11000	or	10110		
12	100010	or	10101	or	11010		
13	100001	or	11001	or	10111		
14	100100	or	100011	or	11100	or	11011
15	101000	or	100110	or	11110		
16	101010	or	100101	or	11101		
17	101001	or	100111	or	11111		
18	1000000	or	110000	or	101100	or	101011
19	1000010	or	110010	or	101110		
20	1000001	or	110001	or	101101		

An investigation of the results in Tables 8.1 and 8.3 shows that although the canonical representation using either Fibonacci or Lucas numbers is the minimal (or one of the minimal) representations in the sense that it contains the minimum number of 1's of all possible representations, it contains the maximum number of digits (including zeros). The result of this observation is that Eqs. (8.5) and (8.10) may be taken as reliable measures of N when determining the canonical representations of numbers.

An extension of these representation theorems to a generalized Fibonacci sequence may seem to be of interest. However, the investigation of such possibilities is unproductive. Completeness requires a set of numbers to include 1 so that the number 1 may be represented. It also requires either 2 or a second 1 so that the number 2 may be represented. Considering only positive seed values for the sequence and noting that 0 is not required for completeness it is easy to see that only three combinations of two seed values are possible; 1 and 1; 1 and 2; and 2 and 1. The first two generate the Fibonacci sequence and the third generates the Lucas sequence.

CHAPTER 9

OPTIMAL SPACING AND SEARCH ALGORITHMS

A number line, say of length 1 unit from [0,1), may be divided into two segments by placing a point on the line at a location determined by the value of a real number, R. The terminology [0,1) indicates a line (or set) which includes the point 0 but excludes the point 1. If $R \geq 1$ then it may be plotted on the line by using the fractional part of R, $fract(R)$. The line may be divided into $n+1$ segments by plotting the n values given by

$$fract(R), \qquad fract(2R),\ldots \qquad fract(nR) \ . \qquad (9.1)$$

If R is a rational number then a limited number of distinct points will be produced by Eq. (9.1) even in the limit of $n \to \infty$. For example, if $R = 0.5$ then only two points will be produced by Eq. (9.1) since $fract(0.5m) = 0$ for all even m and $fract(0.5m) = 0.5$ for all odd m.

An irrational number, on the other hand, will divide the line into $n + 1$ segments. An important theorem relating to this is as follows:

> If the n points generated by $fract(mR)$ for $m = 1,2,3,\ldots,n$ for an irrational value of R are plotted on the line [0,1) then the resulting $n + 1$ line segments will have at most three different lengths.

(The corresponding theorem relating to rational numbers states that at most two different segment lengths occur.) Although a formal proof of this theorem is not obvious, the general validity of this statement can be demonstrated by plotting a few examples for different values of R.

For certain applications it is desirable to have the most uniform division of the number line for a given value of n. That is, the three segments should be as close as possible to the same length for any value of n. The possible applications of this are discussed later in this chapter and in Chapter 13. If R is much less than unity then the procedure given in Eq. (9.1) will yield a cluster of points around zero and one much larger segment for the rest of the line. If R is near unity then the cluster of points will occur near 1.0. If R is near 0.5 then the points will occur near 0, 0.5 and 1.0. In any of these cases the distribution of segment lengths, at least for small n, will not be very uniform. It has been determined that the most uniform distribution of points on the line will occur for $R = \tau$ (or equivalently $R = \tau^k$ for integer k). The first ten points generated using this algorithm are illustrated in Fig. 9.1. In this case it is seen that identically one point occurs in each 0.1 interval on the number line and that three different length line segments, S, M, and L are produced. The ratios of the lengths of these segments can be shown to be

$$\frac{L}{M} = \tau$$

$$\frac{M}{S} = \tau$$

$$\frac{L}{S} = \tau^2 \ . \tag{9.2}$$

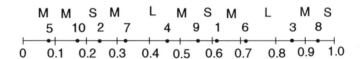

Fig. 9.1. Division of the number line [0,1) by the ten points generated by the function *fract(mτ)* (for $m = 1,2,3,...,10$). Numbers associated with the points give the values of m and indicate the sequence in which the points are generated. The designation S, M and L refer to the three distinct segment lengths, $S \approx 0.0557$, $M \approx 0.0902$ and $L \approx 0.1459$. Note that $L/M = M/S = m\tau$.

Plotting the points given by *fract(mτ²)* (for $m = 1,2,3,...,10$) will yield the same points plotted in the same order as *fract(mτ)*. However, it is straightforward to show that the case for τ^2 is not extended to higher powers. Table 9.1 gives the sequence of points which are plotted first on the number line by *fract(mτ^k)* (i.e. for $m = 1$) as a function of k in terms of those which are plotted for $k = 1$. It is obvious that these numbers follow a Fibonacci sequence.

The same procedure may be used to distribute points on the circumference of a circle. The points may be plotted as a function of angle around the circle as

$$\theta = \mathrm{mod}_{2\pi}\left(\frac{2\pi n}{\tau^k}\right) , \tag{9.3}$$

where $\mathrm{mod}_{2\pi}$ means the remainder of a division by 2π. This is shown in Fig. 9.2 for a value of $k = 1$. The trends as a function of k for this problem are also described by Table 9.1.

Table 9.1. Location of the first point plotted on the line $[0,1)$ as a function of k for the generating function $fract(m\tau^k)$ as described in the text. The points are labeled according to the sequence shown in Fig. 9.1.

k	first point plotted
1	1
2	1
3	2
4	3
5	5
6	8

The division of an interval into segments is directly related to the methods of locating the minimum (or maximum) of a function. The method as described below may be applied if it is known that the function is unimodal on the interval $a < x < d$. Unimodal means that the function has only one minimum (or maximum). Mathematically speaking, this requirement is satisfied if the second derivative of the function does not change sign over the interval. Figure 9.3 shows an example of such a function. The location of the minimum in this function may be found in the following way: Two points b and c are chosen in the interval such that the line segments lengths $ab = cd < ad/2$. The function is evaluated at the points b and c, i.e. $f(b)$ and $f(c)$, respectively. If $f(b) < f(c)$ then the minimum is in the interval ac, if $f(b) > f(c)$ then the minimum is in the interval bd. This can be seen in Fig. 9.3, where the minimum occurs in the interval ac. The new interval, in this case ac, is divided by two new points and the location of the minimum is

$$\theta \longrightarrow$$

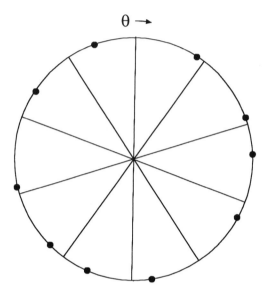

Fig. 9.2.   Division of the circle by the point generated by $\mathrm{mod}_{2\pi}\,(2\pi m/\tau^k)$ (for $k = 1$ and $m = 1,2,3,...,10$). The origin is marked by $\theta = 0$ and the direction of increasing θ is shown by the arrow. The radial lines divide the circle into ten equal segments in order to better illustrate the distribution of points on the circumference.

determined within the new sub-intervals. By continuing applying this method the range of x values containing the minimum may be narrowed and the location of the minimum determined to the required accuracy.   This procedure is illustrated graphically in Fig. 9.4 where the lengths of the line segments are given by λ_i.

It is necessary to consider the means by which the location of the points b and c (and the points which divide the subsequent line segments) are to be determined. Specifically, it is necessary to determine the relative lengths λ_i/λ_{i-1}, $\lambda_{i-1}/\lambda_{i-2},...$ etc. such that the algorithm functions efficiently.   One choice for the relative segment lengths is

$$\frac{\lambda_i}{\lambda_{i-1}} = \tau \tag{9.4}$$

for all i. This algorithm is called the *golden ratio search* and it is a highly efficient method for minimizing a function (Pierre 1986).   Another, somewhat more sophisticated, method is to allow the ratio λ_i/λ_{i-1} to vary as a function of i as

$$\frac{\lambda_i}{\lambda_{i-1}} = \frac{F_i}{F_{i-1}} \tag{9.5}$$

This method is referred to as the *Fibonacci search*. This procedure cannot be continued indefinitely as it will terminate for $i = 2$; i.e. $F_2/F_1 = 1$. This must be taken into account in the initial choice of i. That is, the initial i must be chosen to be large enough to allow for enough iterations in order to obtain the location of the minimum to the desired accuracy.

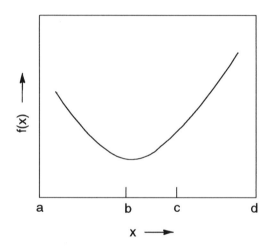

Fig. 9.3. An example of a unimodal function $f(x)$ on the interval $[a,d]$.

It can be shown that the Fibonacci search is more efficient than the golden ratio search; that is, after a given number of iterations the range of possible x values for the location of the minimum is smaller in the former case. It has been suggested that the Fibonacci search is the most efficient search algorithm. Although the proof of this assertion is not clear, it is clear that the Fibonacci search (as well as the golden ratio search) is an efficient method in many instances.

A problem related to the above discussion deals with searching an ordered list of numbers for a known value (or the number closest to the known value). Such an

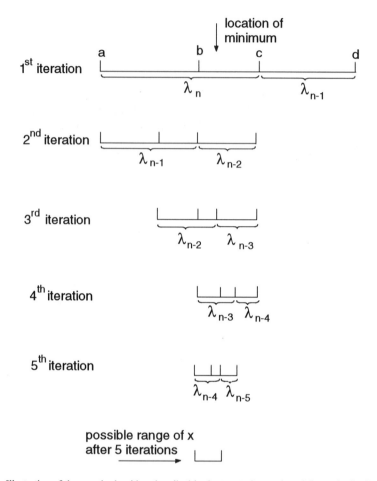

Fig. 9.4. Illustration of the search algorithm described in the text to locate the minimum in the function of Fig. 9.3.

algorithm allows for a value to be inserted into an ordered list in the proper location, or equivalently, for an unordered list to be ordered by inserting each term in its correct place. A straightforward method of searching a list for the purpose of inserting a new term is to divide the list into half and to compare the new term with the value in the middle of the list. Depending on whether the new term is larger or smaller than the value from the list, one or the other half of the list is divided into half and the procedure is repeated. This is continued until the proper location is found. This type of search is referred to as a binomial search. The golden ratio search divides the list into portions with relative numbers of terms given by τ. In certain cases this method is more efficient than the common binomial search. The use of these methods in the development of computer codes can deal, not only with the ordering of lists of numbers, but also with alphabetizing and the location of items in lists of names, addresses or other records and, as such, have numerous and important applications.

CHAPTER 10

COMMENSURATE AND INCOMMENSURATE PROJECTIONS

In Chapter 5 the distinction between a periodic sequence as in Eq. (5.17) and a quasiperiodic sequence as in Eq. (5.16) was discussed. The sequences given in these equations may be displayed in a one dimensional graphical form by representing an A by a long line segment and a b by a shorter line segment as illustrated in Fig. 10.1. If a set of points is used to divide a line into long and short segments in either a periodic arrangement or in a quasiperiodic arrangement then these points form either a periodic or quasiperiodic array (sometimes called a lattice). Thus Figs. 10.1a and 10.1b may be referred to as one dimensional periodic and quasiperiodic lattices, respectively. The simplest kind of periodic array in one dimension is a sequence of evenly spaced points as shown in Fig. 10.2a. This may be viewed as a sequence of line segments of equal length.

One dimensional problems often have two or three dimensional analogues. In the case of a periodic array, the simplest two dimensional analogue is the array or lattice of points described by the vertices of an arrangement of squares as shown in

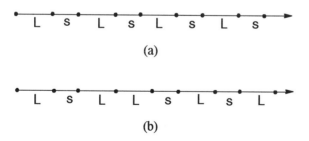

(a)

(b)

Fig. 10.1. (a) A periodic arrangement of long and short line segments in one dimension and (b) a quasiperiodic arrangement of long and short line segments in one dimension. Compare with the sequences given in Eqs. (5.17) and (5.16), respectively.

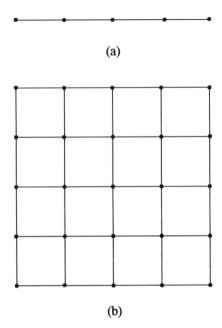

(a)

(b)

Fig. 10.2. (a) One dimensional periodic array formed from a single cell type (line segment) and (b) two dimensional periodic array formed from a single cell type (square).

Fig. 10.2b. This lattice may be extended indefinitely in all (two) dimensions by adding more squares to the array. The method of adding more squares is fairly obvious. New squares are created by translating previous squares by an integral number of edge lengths in directions which are parallel to the edges of the square This process can be repeated indefinitely to produce an infinite square lattice. This is analogous to the discussion in Chapter 5 concerning the generation of an infinite periodic sequence by the repetition of a basic sequence unit. This property of periodic sequences or arrays is referred to as translational symmetry.

More complex periodic lattices in two dimensions can be formed and many of these are discussed in further detail in Chapter 11. As well, three dimensional periodic arrays are also possible. The simplest type of structure is based on the points formed by the vertices of an array of cubes. Again more complex structures are also possible, and Chapter 12 describes some of these.

A quasiperiodic sequence in one dimension cannot be generated using only a single length of line segment; at least two different lengths are needed. The idea of a quasiperiodic lattice (sometimes called a quasilattice) in two or three dimensions is somewhat more obscure than in the one dimensional case. Clearly more than one type of structural unit or cell must be used to produce a quasiperiodic array, just as more than one kind of unit (i.e. rabbit or line segment) is needed in one dimension. It is not obvious, however, whether two different cells will be sufficient or how to go about constructing such an array. It may seem that certain rules for constructing a quasiperiodic array in two or three dimensions which are analogous to the rabbit breeding rule used to generate the one dimensional array would be needed. This approach is possible, although not as straightforward as in the one dimensional case, and will be discussed in detail in Chapter 11. Another approach involves the so-called cut and projection method. This method is the topic of the present chapter although here the discussion will concentrate on the application of this method for generating one dimensional structures and extensions to higher dimensions will be dealt with later. The basic principles of this method are as follow.

A two dimensional periodic lattice is constructed as shown in Fig. 10.3. In this

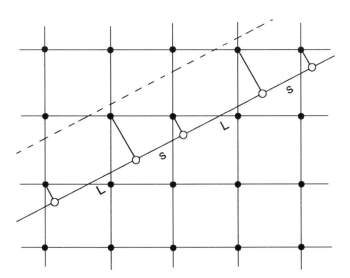

Fig. 10.3. Cut and projection from two dimensions to one dimension with a cut angle with a rational tangent (tan $\theta = 1/2$). This is the so-called commensurate cut and projection which yields the periodic array as shown in Fig. 10.1a. The cut line is shown by the solid line and the band is defined by the broken line.

case the lattice is constructed from squares, although the same methods may be applied to any periodic array. A cut line is constructed which intersects the two dimensional lattice as shown in the figure. A band is constructed on one side of the cut line which includes the lattice points nearest to the line. These points are projected onto the line by constructing a line through the lattice point which is perpendicular to the cut line. This procedure will yield an array of points on the cut line which are formed by the intersections of the projection lines as shown in Fig. 10.3. The line segments which are created by this procedure are either long (L in the figure) or short (s in the figure). These long and short segments form an array and the details of the array are determined by the angle, θ, which the cut line makes with the edges of the squares in the two dimensional lattice. In the example shown in Fig. 10.3 the cut line is characterized by an angle with $\tan \theta = 1/2$. It is significant that the tangent of the cut angle in this example is a rational number and the procedure described here is referred to as a commensurate cut and projection. A commensurate cut and projection from a periodic array will yield a periodic array in a lower dimension. In the present example the cut and projection from the two dimensional square lattice will yield a periodic array in one dimension. In fact close examination of the sequence of L and s segments formed on the cut line in Fig. 10.3 will reveal that the arrangement of L and s is identical to the arrangement of A and b in Eq. (5.17) and the arrangement of long and short line segments in Fig. 10.1a. It is of relevance to note that the tangent of the cut angle in this example is 1/2. This is the characteristic of the fraction of the line segments which are long, i.e. the ratio of the number of long segments to the total number of segments is 1/2.

Changing the cut angle will result in different arrangements of long and short line segments in the projected sequence on the cut line. As long as the tangent of the cut angle is a rational number, that is the ratio of two integers, then the ratio of the number of long and short line segments in the sequence will be a rational number and the resulting pattern will be periodic. However, if the tangent of the cut angle is an irrational number then the cut is said to be incommensurate and the ratio of the long to short line segments in the projected sequence will also be an irrational number. Clearly in this case the sequence cannot be periodic as the true ratio of long to short segments can only be exhibited for an infinite sequence. An example of this type of cut and projection is illustrated in Fig. 10.4. A cut angle with a tangent which is related to the golden ratio is of most relevance to the present discussion and the example shown in the figure is for the case $\tan \theta = 1/\tau$.

As seen in the figure, the sequence of long and short line segments as projected onto the cut line is given by

$$LsLLsLsLLsLLs...\ .\qquad(10.1)$$

It is easy to see that this is identical to the quasiperiodic sequence given for the adult and baby rabbits in Eq. (5.16). On the basis of the discussion above for the commensurate projection, it is expected that the ratio of the number of long segments to the total number of segments should be $1/\tau$. This is the ratio which is characteristic of the Fibonacci sequence and it is, therefore, not surprising that the

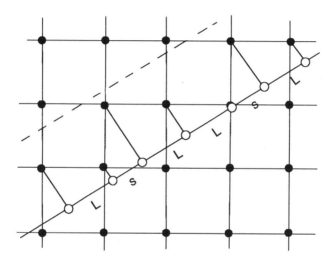

Fig. 10.4. Cut and projection from 2 dimensions to one dimension with a cut angle with an irrational tangent (tan $\theta = 1/\tau$). This is the so-called incommensurate cut and projection which yields the quasiperiodic array as shown in Fig. 10.1b. The cut line is shown by the solid line and the band is defined by the broken line.

arrangement of long and short line segments which results from a cut and projection at an angle with tan $\theta = 1/\tau$ would yield this kind of quasiperiodic sequence. It can also be shown from a simple geometric analysis of the problem that the relative lengths of the long and short segments on the cut line are in the ratio $\tau : 1$. As the sequence of long and short line segments displayed by the projection of lattice points onto the cut line is characteristic of a Fibonacci sequence the resulting sequence of points in one dimension is generally referred to as a Fibonacci lattice. This concept can be extended to two or three dimensions and will be discussed further in the next two chapters.

The idea of rational approximants, as discussed in Chapter 7, can be nicely demonstrated by the use of the cut and projection method. It is clear that the ratio of successive Fibonacci numbers F_{n+1}/F_n approaches the golden ratio in the limit of large n. It is similarly apparent that the ratio F_n/F_{n+1} should approach $1/\tau$ in the limit of large n. In the above examples cut angles with tangents which are representative of two possible ratios of Fibonacci numbers have been presented; in one case $1/2 = F_2/F_3$ and $1/\tau = F_n/F_{n+1}$ in the limit of $n \to \infty$. Values of tan θ which are equal to other ratios of Fibonacci numbers can also be considered. Table 10.1 gives some examples. Each cut and projection will yield a characteristic sequence of long and short line segments. In the case of the projection at a cut angle with tan $\theta = 1/2$ the sequence was seen to consist of the elements $LsLsLs...$, or a repetition of the basic unit or cell Ls. It is easy to see that a cut at an angle with tan $\theta = 1/1$ (or a cut angle of 45°) will yield projected line segments on the cut line which are all the same length. This is represented by the sequence $LLLL...$ or a repetition of the basic cell L. Table 10.1 shows the basic cell which is obtained (as well as its length) for cut and projections at angles with tangents equal to other ratios of Fibonacci numbers. It is seen that the basic cell is a Fibonacci sequence in each case with a length that increases (as the sequence of Fibonacci numbers) as the value of tan θ is given by the increasing rational approximants of the inverse of

Table 10.1. Sequences of long and short line segments produced by the cut and projection method from two dimensions to one dimension for cut angles whose tangents are ratios of successive Fibonacci numbers, F_n/F_{n+1}.

n	tan $\theta = F_n/F_{n+1}$	basic cell	cell length
1	1/1	L	1
2	1/2	Ls	2
3	2/3	LsL	3
4	3/5	$LsLLs$	5
5	5/8	$LsLLsLsL$	8
6	8/13	$LsLLsLsLLsLLs$	13
7	13/21	$LsLLsLsLLsLLsLsLLsLsL$	21

the golden ratio, $1/\tau$. In the limit of $n \rightarrow \infty$, the rational approximant approaches the true irrational value of $1/\tau$ and the basic cell becomes the infinite quasiperiodic Fibonacci sequence.

The Fibonacci sequence produced by the cut and projection at an angle with $\tan \theta = 1/\tau$ can be used to demonstrate the concept of deflation (or inflation). The simplest way of demonstrating this behavior is by analogy with the deflation of the Fibonacci rabbit sequence described in Chapter 5. Since the sequence shown in Fig. 10.1b is identical to the sequence of Fibonacci rabbits of Fig. 5.6 if adult rabbits (A) are represented by long line segments (L) and baby rabbits (b) are represented by short line segments (s), then a rescaling using the same rules as was applied to the deflation of the Fibonacci rabbit sequence should rescale the one dimensional Fibonacci lattice. In the present case these deflation rules are expressed as follows:

(1) remove all isolated L's from the sequence
(2) replace all LL's by s's and
(3) replace all original s's by L's.

This procedure is shown in Fig. 10.5. A rescaling of the one dimensional Fibonacci lattice to shorter lengths, a procedure which is referred to as inflation (by analogy with the discussion in Chapter 3), can also be accomplished. This is shown in Fig. 10.6. Each long line segment is divided into two segments with length ratios of τ:1 with the longer segment on the left. It is easy to see that the length of the longer of these newly formed segments in the same as the length of the original short segment. This inflation process is, therefore equivalent to the rescaling procedure;

(1) replace all s by L
(2) replace all L by Ls.

This procedure can be applied to the Fibonacci rabbit sequence as well by substituting A for L and b for s. For the one dimensional Fibonacci lattice this procedure is illustrated in Fig. 10.6. Since the ratio of the number of long line segments to total line segments in a Fibonacci lattice must be $1/\tau$, then this ratio should not change during rescaling. It is easy to show that this is the case. The number of original long segments is related to the number of original short segments by

$$N_L = \tau N_s \ . \tag{10.2}$$

The number of new long segments is given by the number of original short segments, N_S, plus the number of new long segments, which is equal to the number of original long segments;

$$N_L' = N_s + N_L \qquad\qquad (10.3)$$

and from Eq. (10.2)

$$N'_L = N_s(1+\tau) \ . \qquad\qquad (10.4)$$

The number of new short line segments is merely the number of original long segments;

$$N_s' = N_L \ . \qquad\qquad (10.5)$$

Equations (10.4) and (10.5), therefore, give the ratio of new long to new short segments as

$$\frac{N_L'}{N_s'} = \frac{N_s(1+\tau)}{N_L} = \frac{(1+\tau)}{\tau} = \tau \ . \qquad\qquad (10.6)$$

This demonstrates that the rescaled Fibonacci lattice is also a valid Fibonacci lattice. This procedure, as shown in Fig. 10.6, can be repeated indefinitely.

The properties described here for the one dimensional Fibonacci lattice will be extended to higher dimensions in the next two chapters.

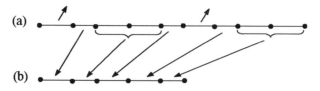

Fig. 10.5. One dimensional Fibonacci lattice; (a) original lattice and (b) lattice after one deflation as described in the text.

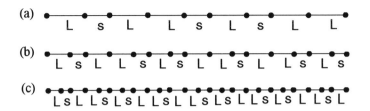

Fig. 10.6. One dimensional Fibonacci lattice; (a) original lattice, (b) after one inflation and (c) after two inflations.

CHAPTER 11

PENROSE TILINGS

The ideas concerning the generation of quasiperiodic structures as described in the previous chapter may be extended to higher dimensions. In the present chapter the concept of a two dimensional Fibonacci lattice is explored. Historically speaking, the first detailed work on two dimensional Fibonacci lattices was published in 1974 by Roger Penrose (1974), although certain ideas concerning these types of structures have been in existence for many years. It was, however, Penrose who first studied and properly appreciated the mathematical concepts present in these two dimensional Fibonacci lattices and it is, therefore, appropriate that they are generally referred to as Penrose tilings. In general the word *tiling* is used to describe a structure which is comprised of one or more types of cells or tiles which can be used to fill space. It is most common to think of a tiling in two dimensions, such as the arrangement of floor tiles, although in a mathematical sense the concept is valid in any number of dimensions. In this case the tiling consists of a finite number of different two dimensional figures which can be used to cover a plane without overlap. As expected, the Penrose tilings exhibit quasiperiodic rather than periodic order. There is extensive literature on both periodic and quasiperiodic tilings and many of the properties of these structures are beyond the scope of the present book and the present discussion will cover only those topics which are of relevance to the golden ratio. However, a basic introduction to the mathematics of tilings may be found in Grundbaum and Shepherd (1987) and in Dunlap (1990).

Perhaps the most straightforward method of constructing a Penrose tiling is by analogy with the cut and projection method described in the previous chapter for the construction of a one dimensional Fibonacci lattice. An incommensurate cut and projection from a two dimensional periodic (e.g. square) lattice at a cut angle with a tangent related to the golden ratio will yield a one dimensional Fibonacci lattice. Analogously an incommensurate cut and projection from a four

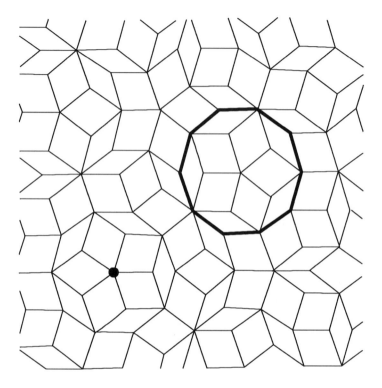

Fig. 11.1. A two dimensional Penrose tiling based on oblate (fat) and prolate (thin) rhombuses. The circle indicates a point of local fivefold rotational symmetry and the heavy lines indicate the ten different tile edge directions. These points are discussed in the next chapter.

dimensional periodic (e.g. hypercubic) lattice at a cut angle with a tangent related to the golden ratio will yield a two dimensional Fibonacci lattice which is usually referred to as a two dimensional Penrose tiling. In this case the cut *line* is actually a plane and is sometimes referred to as a hyperplane for reasons which will become apparent in the next chapter. It is not possible to draw (in two dimensions) the four dimensional hypercubic structure but it is possible to analytically compute and draw the resulting two dimensional Penrose tiling and this is shown in Fig. 11.1. There are several interesting properties of the two dimensional Penrose tiling shown in the figure.

(1) The tiling is made up of two different tile shapes (or cells); one is an oblate (or

fat) rhombus and the other a prolate (or thin) rhombus. The edges of these two tiles are all of the same length and the angular relationships are illustrated in Fig. 11.2.

(2) Certain points in the tiling are surrounded by a star-like arrangement of rhombuses and possess fivefold point symmetry. This feature is referred to as rotational symmetry.

(3) The edges of the rhombuses are oriented in a number of well defined directions (in this case there are ten different directions). This feature is referred to as bond orientational order.

(4) The tiling cannot be extended by translating and copying any section of the tiling (as is the case for the two dimensional square lattice in Fig. 10.2b).

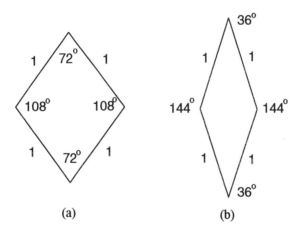

$$(a) \qquad\qquad (b)$$

Fig. 11.2. (a) Oblate rhombus and (b) prolate rhombus from the Penrose tiling of Fig. 11.1 showing the angle and edge length relationships.

These features are characteristic of quasiperiodic (sometimes called aperiodic) order in two dimensions. This situation is completely analogous to the one dimensional case of the Fibonacci sequence as shown in Fig. 5.6 or Fig. 10.1b. In particular some of the similarities between the one and two dimensional cases which are of particular importance are as follows:

(1) The ratio of the numbers of the two different tile shapes as the size of the tiling goes to infinity is related to the golden ratio and is given by

$$\frac{N_{oblate}}{N_{prolate}} = \tau \tag{11.1}$$

(2) The quasiperiodic nature of the tiling is not a direct result of the tile shapes but is a consequence of the rules for arranging the tiles. Thus, as in the case of Fig. 10.1a, the two tile shapes can be arranged in a periodic lattice as well. For the two dimensional rhombuses of Fig. 11.1, one example of a periodic tiling which can be constructed is shown in Fig. 11.3.

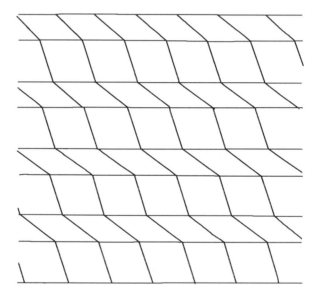

Fig. 11.3. A two dimensional periodic tiling (or lattice) comprised of the oblate and prolate rhombuses of Fig. 11.1.

The above observations clearly illustrate the relationship of the two dimensional Penrose tiling and the golden ratio. This relationship is also apparent from the geometry of the Penrose rhombuses themselves. As Fig. 11.4 shows each rhombus can be easily dissected into two golden triangles. This feature will be discussed further below in the context of deflation operations.

It was seen that the one dimensional Fibonacci lattice could be generated in two ways; by the incommensurate cut and projection method and by appropriate generation rules (in this case the rabbit breeding rules). In a more general sense the rules used to generate the Fibonacci sequence can be referred to as matching rules. An analogy of the generation of the Fibonacci lattice by matching rules also exists for the two dimensional Penrose tiling. In fact it was through the use of matching rules (rather than by the cut and projection method) than Penrose first

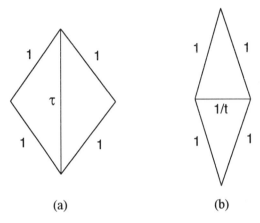

(a)   (b)

Fig. 11.4. (a) Dissection of the oblate Penrose rhombus into two obtuse golden triangles (golden gnomons) and (b) the dissection of the prolate Penrose rhombus into two acute golden triangles.

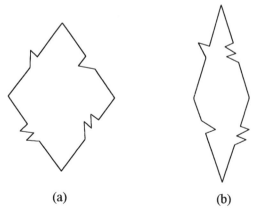

(a)   (b)

Fig. 11.5. Matching rules in the form of keys on the (a) oblate and (b) prolate rhombuses of the Penrose tiling in Fig. 11.1.

proposed the existence of the two dimensional analogy to the Fibonacci sequence. The matching rules for the two dimensional tiling can be expressed in the form of keys on the tiles; rather like the shape of the pieces of a jigsaw puzzle. For the tiling of Fig. 11.1 the matching rules are illustrated in Fig. 11.5. These keys will allow the tiles to be arranged as they are in Fig. 11.1 but not as they are in Fig. 11.3. The use of the matching rules for the placement of the tiles is a necessary condition for the generation of a proper quasiperiodic tiling in two dimensions (that is a correct two dimensional analog of the Fibonacci lattice in one dimension). However, the use of the matching rules, in itself, is not sufficient to guarantee that the resulting tiling will be a correct Fibonacci lattice. Thus the construction of the two dimensional Fibonacci lattice is not so straightforward as the one dimensional case which could be produced precisely from the rabbit breeding rules.

The Penrose tiling shown in Fig. 11.1 is not the only two dimensional quasiperiodic lattice which exhibits the properties described above. In fact, the first quasiperiodic tiling proposed by Penrose is shown in Fig. 11.6. This earlier tiling is distinct from those discussed previously in this book as it is based on four tile shapes instead of two. However, there are certain similarities between these different tilings; i.e. the existence of fivefold rotational symmetry and a construction based on matching rules.

Another quasiperiodic tiling of interest is illustrated in Fig. 11.7. The tile shapes here are referred to as darts and kites. The occurrence of fivefold symmetry at specific points in the structure is obvious in the figure. It can be shown that these two tile shapes occur in the ratio

$$\frac{N_{kites}}{N_{darts}} = \tau \quad . \tag{11.2}$$

The matching rules for these tiles which are necessary for the construction of a proper quasiperiodic tiling are illustrated in Fig. 11.8. Again, as in the case of the tiling based on rhombuses, the matching rules are a necessary but not sufficient condition to yield a true two dimensional Fibonacci lattice. It is not surprising that the tiles themselves used in the production of this tiling exhibit properties which are related to the golden ratio. In fact it is readily seen from the geometry of the tiles as shown in Fig. 11.9, that both tile shapes can be dissected into golden triangles.

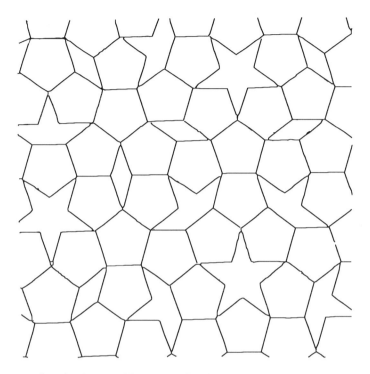

Fig. 11.6. A two dimensional Penrose tiling based on four tile shapes (pentagons, rhombuses, three and five pointed stars) as suggested by Penrose. Adapted from Penrose (1974).

The idea of rescaling a Fibonacci lattice in two dimensions (i.e. deflation) which has been discussed at length with respect to the one dimensional Fibonacci lattice, can be demonstrated here as well. In fact, this property follows directly from the fact that the tilings utilized in these two dimensional Fibonacci lattices are composed of golden triangles and a rescaling of the golden triangles by a factor of the golden ratio has been shown in Fig. 3.3.

A simple example of rescaling the tiling of Fig. 11.1 is illustrated in Fig. 11.10. This procedure may be repeated indefinitely, scaling the tiling to increasing or decreasing tile size, while still maintaining the correct ratio of oblate to prolate rhombuses to produce a two dimensional Fibonacci lattice. Improper tilings are characterized by tiling *mistakes*. Although these mistakes are not necessarily obvious immediately they will result in an inconsistency in the tiling which will

Fig. 11.7. Two dimensional Penrose tiling comprised of darts and kites.

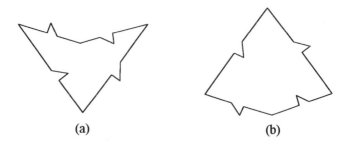

(a)   (b)

Fig. 11.8. Matching rules in the forms of keys on the (a) darts and (b) kites of the tiling in Fig. 11.7.

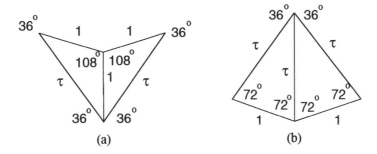

Fig. 11.9. Dissection of darts and kites, (a) and (b), respectively, into golden triangles. Angular and length relationships are illustrated.

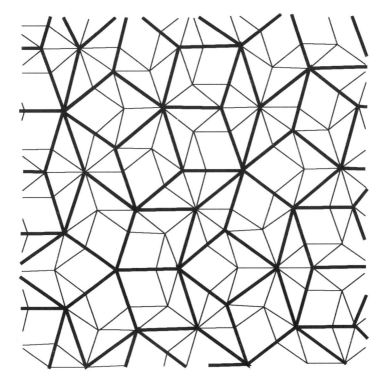

Fig. 11.10. Quasiperiodic tiling in two dimensions illustrating the application of a deflation operation.

eventually become evident if the tiling is extended. This will create a situation where a tile cannot be added to the lattice in such a way as to be consistent with the matching rules. Following an analogy with the one dimensional case, tiling mistakes, as given in Table 5.6, will become obvious if the relevant section of the tiling is deflated repeatedly.

There are several historical aspects of two dimensional Fibonacci tilings which are of particular interest. Although it was Roger Penrose who first appreciated the mathematical importance of quasiperiodic tilings, similar patterns have been considered previously. A few examples are given.

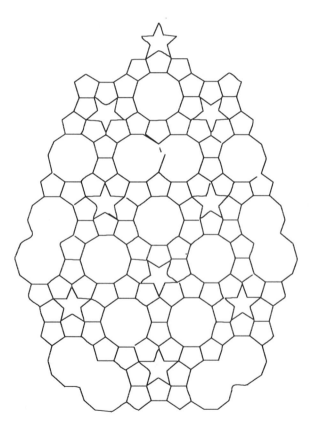

Fig. 11.11. Tiling with fivefold symmetry designed by Kepler (1619).

Johannes Kepler (1571-1630) had an interest in a variety of fields of natural science. A topic which he investigated in considerable depth concerned the relationship of different geometrical forms. Part of this study dealt with the construction of two dimensional tilings. One tiling presented by Kepler in his monograph of 1619, *Harmonices Mundi* (Kepler 1619) is illustrated in Fig. 11.11. This tiling is composed of four tile shapes, regular pentagons, pentagrams (five pointed stars), decagons and double decagons and shows certain similarities with the quasiperiodic tiling in Fig. 11.6. In fact it can be shown that the tiling of Fig. 11.11 may be extended indefinitely to produce a proper quasiperiodic tiling.

Prior to Kepler's investigations, the German artist and scientist Albrecht Dürer (1471-1528) considered tilings which are of relevance to the present discussion. Fig. 11.12 shows some examples of tiling published by Dürer around 1525 (Dürer 1525). These tilings illustrate Dürer's interest in the regular pentagon as well as his recognition that pentagons and rhombuses could be arranged periodically as in Fig. 11.12a or quasiperiodically as in Fig. 11.12b. A detailed analysis of the contribution of Dürer to an understanding of fivefold symmetry has been given by Crowe (1992).

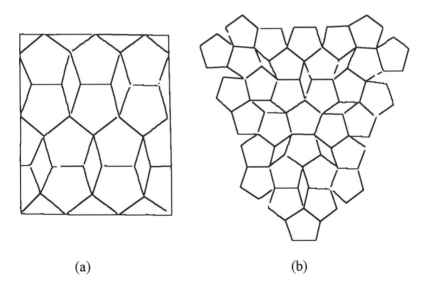

(a)   (b)

Fig. 11.12. Tilings with fivefold symmetry designed by Dürer (1525) which exhibit (a) periodic order and (b) no periodic order.

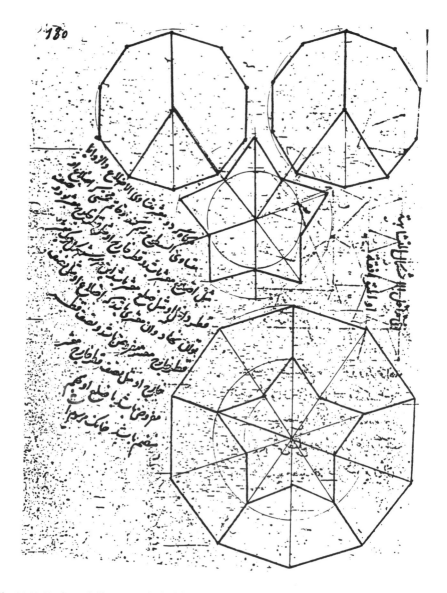

Fig. 11.13. Portions of tilings comprised of darts and kites designed by Abu'l Wafa'al Buzjani (c. 1180). From Chorbachi and Loeb (1992), copyright World Scientific Publishing.

Perhaps the earliest consideration of tiling which exhibits some properties of quasiperiodicity is in the work of the 12th century Persian mathematician Abu'l Wafa'al Buzjani (c. 1180) who considered the properties of figures with fivefold symmetry. Figure 11.13 illustrates a construction reported by Buzjani. The relationship of darts and kites and their arrangement into figures with fivefold symmetry is clearly evident in this figure and the relationship with quasiperiodic tilings is apparent by a comparison with Fig. 11.7. A detailed analysis of ancient Persian work related to fivefold symmetry has been given by Chorbachi and Loeb (1992).

The above discussion illustrates the geometric constituent necessary to produce a quasiperiodic tiling in two dimensions have been considered since antiquity. However, it was not till the work of Penrose in the early 1970's, that an understanding of these tiling in the sense of two dimensional Fibonacci lattices appeared. This led to an extension of these ideas to three dimensions and an understanding of the structure of quasicrystals as discussed in the next chapter.

CHAPTER 12

QUASICRYSTALLOGRAPHY

An extension of the concepts discussed in the previous chapter to three dimensions describes the unique structure of a newly discovered class of metallic materials. Before discussing the structure of these novel materials, it is useful to consider the models for describing the microscopic structure of atoms in conventional materials.

In a crystalline material the atoms are arranged on a lattice. This is, as has been discussed in Chapter 10, a periodic array of cells of a single type. The fundamental cell used for generating the lattice is called the *unit cell*. In most crystals this is a three dimensional structure and the simplest lattice is an arrangement of cubes. For purposes of illustrating some basic crystal properties it is convenient to consider some two dimensional examples. In two dimensions the simplest lattice is a square lattice as shown in Fig. 12.1. The structure is formed by translating and copying the basic unit cell, the square in order to fill all of two dimensional space. This lattice is, therefore, called a *space filling* structure. Space filling structures in two dimensions may be generated from other cell shapes. Some examples of this are illustrated in Fig. 12.1. In order to form an actual crystal the atoms must be placed on the lattice. The way in which this is done is called the *basis*. The simplest basis consists of a single type atom placed at each lattice point as shown in Fig. 12.2a. It is possible as well to produce more complex structures as shown in Fig. 12.2b where more than a single atom is associated with each lattice point. This construction may be expressed as

$$LATTICE + BASIS = CRYSTAL \ . \qquad (12.1)$$

The lattice and the crystal related to it have some basic symmetry properties. These are

(1) translational symmetry and
(2) rotational symmetry.

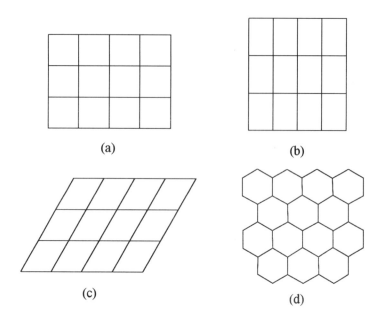

Fig. 12.1. Some examples of lattices (space filling structures) in two dimensions consisting of cells which
         are (a) squares, (b) rectangles, (c) parallelograms and (d) hexagons.

The first property is a direct result of the manner in which the lattice was
generated. The second property results from the basic symmetry properties of the
unit cell. For example the square may be rotated by 90°, 180°, 270° or 360°
without changing its appearance. The same is true of the square lattice (in two
dimensions). Thus the square lattice is said to possess *fourfold* rotational
symmetry (since it can be rotated by multiples of 360/4 degrees). The square lattice
also has twofold symmetry since rotations of 360/2 degrees leave its appearance
unchanged. It is a general feature that lattices with n-fold symmetry also have $n/2$-
fold symmetry when n is even. An inspection of the lattices shown in Fig. 12.1
shows the existence of two twofold symmetry (all lattices), threefold symmetry
(hexagonal lattice), fourfold symmetry (square lattice) and sixfold symmetry
(hexagonal lattice). Although these lattices are not the only ones possible in two
dimensions, the symmetries shown are the only ones which are possible (i.e.
allowed) in two dimensions.

Since the golden ratio has been shown to be related to geometric figures which
exhibit fivefold symmetry, it is interesting to consider the possibility of a lattice

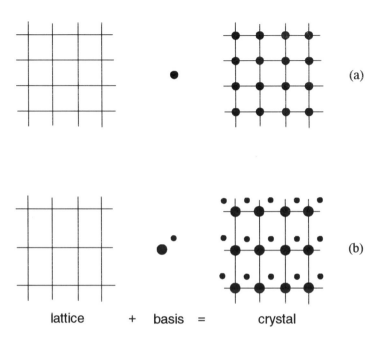

lattice   +   basis   =   crystal

Fig. 12.2. Crystals in two dimensions formed by a lattice and a basis; (a) square lattice with single atom basis and (b) rectangular lattice with two atom basis.

with fivefold symmetry by constructing (or trying to construct) a lattice from cells with fivefold symmetry. Figure 12.3 shows that the construction of a space filling structure comprised of regular pentagons is not possible. Traditionally this demonstration has been taken as evidence that crystallographic fivefold symmetry is not allowed. However, it has been shown in the previous chapter that the two dimensional Penrose tiling has certain features which are characteristic of fivefold symmetry. Figure 11.1, for example, shows that points exist within the quasilattice which have local fivefold rotational symmetry. This structure cannot be produced by translating and reproducing some group of cells. The fact that the correct ratio of tile types is the golden ratio, an irrational number, indicates that this is the case. The quasilattice can, therefore, be said to possess fivefold rotational symmetry but no translational symmetry.

An extension of these ideas to three dimensions is possible. The simplest three dimensional lattice consists of cubic unit cells. This lattice (and the

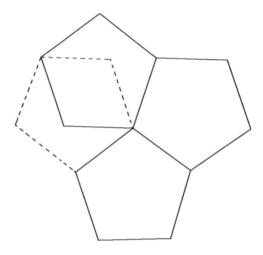

Fig. 12.3. Attempt to construct a two dimensional lattice from regular pentagons showing that adjacent cells overlap.

associated crystal) will have translational symmetry and the rotational symmetry of the cube. Although fourfold symmetry is perhaps the most apparent symmetry of the cube it is not the only symmetry present. The cube shows fourfold (and also twofold) symmetry when viewed along a direction perpendicular to a face. When viewed along an edge direction the cube exhibits only twofold symmetry and when viewed along a vertex direction the cube shows threefold symmetry. These properties are shown in Fig. 12.4.

A similar consideration of the symmetry characteristics of the other Platonic solids is of interest at this point. These are described in Table 12.1. It is obvious from these data that the edges of solids have twofold symmetry. This is because an edge can only be formed from the intersection of two faces. The relationship of the symmetry characteristics of the edges and faces to the values of m (the number of faces per vertex) and n (the number of edges per face) given in Table 4.1 is clear. The solids which exhibit fivefold symmetry, i.e. the dodecahedron and the icosahedron, can be shown to be unsuitable for the formation of a space filling structure (i.e. lattice) in three dimensions. This has traditionally been used as an argument against the existence of three dimensional fivefold symmetry.

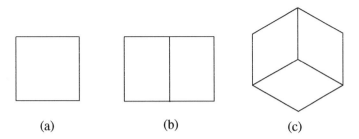

(a) (b) (c)

Fig. 12.4. Symmetry characteristics of the cube; (a) fourfold symmetry axis about the face direction, (b) twofold symmetry axis about the edge direction and (c) threefold symmetry axis about the vertex direction.

Table 12.1. Symmetry characteristics of the Platonic solids. The numbers refer to the rotational symmetry of the solid as viewed from different directions.

solid	vertex	face	edge
tetrahedron	3	3	2
cube	3	4	2
octahedron	4	3	2
dodecahedron	3	5	2
icosahedron	5	3	2

Not all solid materials exhibit crystalline order. Amorphous, or glassy, materials have been known for many years. These materials have atoms which are arranged randomly within the material. The distinction between a crystalline material and an amorphous material is rather like the difference between stacking marbles neatly in a box and throwing them in at random. Because the atoms in an amorphous material are randomly arranged there is no clearly defined unit cell and hence, no translational or rotational symmetry.

A natural extension of the previous discussion concerning Penrose tilings is to consider the possibilities of real three dimensional materials with quasiperiodic order (Penrose 1989). This is particularly interesting as the two dimensional Penrose tiling discussed in Chapter 11 exhibited some aspects of fivefold symmetry. In fact since the work by Penrose in the early 1970's there has been speculation concerning materials with quasiperiodic order. These materials are known as quasicrystals and experimental evidence for their existence was presented in 1984. The structure of three dimensional quasicrystals may be described on the basis of a three dimensional quasilattice which may be generated by an incommensurate cut and projection from a periodic lattice in six dimensions to three dimensions. The simplest case would be a cut and projection from a six dimensional hypercubic structure onto a three dimensional space (called a three dimensional hyperplane).

If the tangent of the irrational cut angle is a power of the golden ratio then the resulting quasiperiodic three dimensional structure will have certain characteristics which are analogous to the properties of the two dimensional Penrose tiling previously discussed. These include;

(1) the existence of two rhombohedral cells (or three dimensional tiles) as illustrated in Fig. 12.5

(2) a ratio of the number of the two tile shapes which is the golden ratio

(3) the formation of localized points in space which are formed by the clusters of tiles which have icosahedral symmetry

(4) the existence of quasiperiodic order.

The features listed above are consistent with the true three dimensional quasiperiodicity characteristic of a three dimensional Penrose tiling or Fibonacci lattice. The tiles, or cells in the three dimensional case are rhombohedra as shown in Fig. 12.5. Each rhombohedron has six rhombuses as shown in Fig. 12.6 as faces. This two dimensional figure is referred to as the *golden rhombus* and has perpendicular diagonals which have a ratio of lengths of $1 : \tau$. The arrangement of the rhombuses determines the geometry of the resulting rhombohedral tiles. The oblate rhombohedron is formed in such a way that two opposite vertices are formed by the intersection of three rhombuses at their obtuse angles. The prolate rhombohedron is formed in such a way that two opposite vertices are formed by the intersection of three rhombuses at their acute angles.

In the same way that a two dimensional Penrose tiling may be constructed from rhombuses using appropriate matching rules, similar rules in the form of *keys* on the faces of the rhombohedral tiles may be used to construct a three dimensional Penrose tiling. Again the keys are a necessary but not sufficient condition for generating a correct Penrose tiling. As in two dimensions, tiling mistakes prevent the quasiperiodic lattice from being extended indefinitely and become apparent during deflation operations.

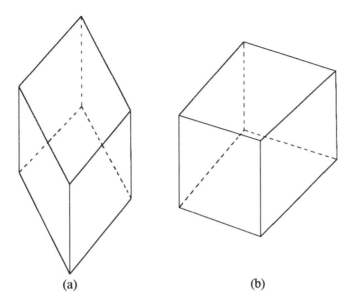

(a)   (b)

Fig. 12.5. Rhombohedral cells formed by a cut and projection from six dimensions to three dimensions; (a) prolate rhombohedron and (b) oblate rhombohedron.

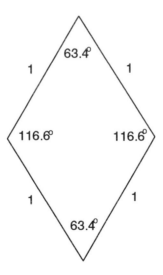

Fig. 12.6. The golden rhombus.

An inspection of the clusters of rhombohedra in the three dimensional Penrose tiling will reveal the existence of local icosahedral rotational symmetry. Insight into the symmetry of the three dimensional Penrose tiling can be gained from a consideration of the properties of the six dimensional hypercubic structure. In a three dimensional cubic structure the cube is formed by vectors along three orthogonal (or mutually perpendicular) axes. In a six dimensional space the hypercube is formed by vectors along six orthogonal axes. This construction is not possible in three dimensions. However, the construction of six equally spaced vectors in three dimensions will define the six principal directions for the icosahedron as illustrated in Fig. 12.7. Thus the symmetry characteristics of the six dimensional hypercube are similar to those of the three dimensional icosahedron which is derived from it by the cut and projection method. It is of interest that the angle between adjacent vertex directions for the icosahedron is the apex angle of

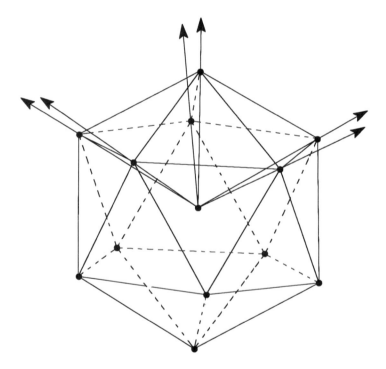

Fig. 12.7. Principal axes of the icosahedron defined by the directions from the center to each of the vertices in the top half of the solid. The directions to the vertices in the lower half of the solid are merely extensions of the lines already shown.

the golden rhombus, $63.435°$. Thus the triangle formed between two icosahedron vertex directions and one icosahedron edge is precisely one half of the golden rhombus.

A large number of materials are now known to possess a structure based on the three dimensional Penrose tiling and are commonly referred to as quasicrystals. As is the case with conventional crystalline materials, it is necessary to specify the arrangements of the atoms relative to the lattice points. In the case of quasicrystalline materials this is expressed as

$$QUASILATTICE + DECORATION = QUASICRYSTAL . \quad (12.2)$$

This concept is the same as that expressed in Eq. (12.1) although the customary terminology is different.

In actual quasicrystalline structures the atoms are most commonly placed at vertex points in the quasilattice or along lines between the vertices. Thus the edges of the rhombohedra in the three dimensional Penrose tiling define the directions of the chemical bonds between atoms. For this reason it is sometimes said that these materials exhibit *bond orientational ordering*. This can be readily seen in the two dimensional case by an inspection of the tiling in Fig. 11.1 where it is seen that precisely ten different rhombus edge directions are possible.

In a manner analogous to the crystalline case, a quasicrystalline material exhibits rotational symmetry which is characteristic of the point symmetry of the quasilattice. In most quasicrystalline materials this symmetry is the symmetry of the icosahedron. These materials are, therefore, referred to as icosahedral quasicrystals. The symmetry characteristics of the icosahedron are observed along the vertex, face and edge directions as illustrated in Fig. 12.8. This figure shows the existence of five-, three- and twofold rotational symmetry along these directions, respectively.

Probably the most straightforward method of observing the symmetry characteristics of a material is by means of a diffraction experiment using x-rays or electrons. The diffraction pattern obtained in this way will have the symmetry of the structure as viewed along the direction of the incident x-rays or electrons. Results of an electron diffraction experiment on quasicrystalline $Al_{65}Cu_{20}Fe_{15}$ are shown in Fig. 12.9. These measurements have been made along directions which have been determined by the angular relationships of the vertex, face and edge directions of an icosahedron. The patterns observed in the figure demonstrate the five- three- and twofold symmetry expected along these different crystallographic directions.

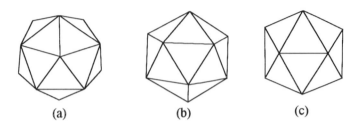

Fig. 12.8. Symmetry characteristics of the icosahedron; (a) fivefold symmetry axis about the vertex direction, (b) threefold symmetry axis about the face direction and (c) twofold symmetry axis about the edge direction.

The idea of rational approximants is related to the formation of novel tilings. In the cut and projection from two dimensions to one dimension, a rational cut angle yields a periodic one dimensional lattice and an irrational cut angle yields a quasiperiodic lattice.   Choosing a cut angle whose tangent is a rational approximant (or the inverse of a rational approximant) of the golden ratio will produce a periodic lattice which is comprised of portions of a Fibonacci lattice. This was demonstrated in Chapter 10.   As the rational approximant becomes closer to the golden ratio, the portion of the Fibonacci sequence which repeats becomes larger.   The same idea can be extended to two and three dimensional lattices.   Clusters of tiles which are portions of the two or three dimensional Penrose tiling can be repeated to produce a tiling which overall is periodic, and therefore possesses translational symmetry, but which has local symmetry characteristic of the quasiperiodic structure.   As the rational approximant defining the tangent of the cut angle becomes closer to the golden ratio the size of the unit cell becomes larger although some overall conventional (i.e. cubic) translational symmetry is maintained.   Such rational approximants of quasicrystalline materials are observed in real systems and are distinguished by the rational approximate defining the cut angle.   Therefore, 1/1, 1/2, 2/3, 3/5, ... etc. rational approximants of the quasicrystalline material may exist.

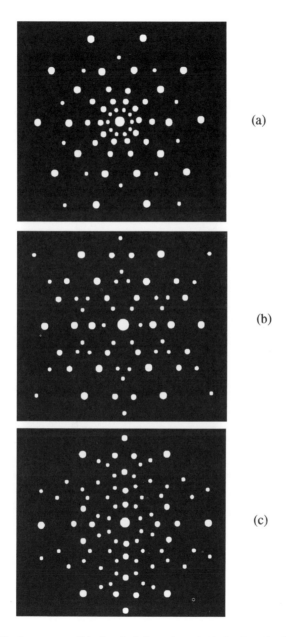

Fig. 12.9. Electron diffraction patterns of the icosahedral quasicrystal $Al_{65}Cu_{20}Fe_{15}$ showing (a) fivefold symmetry axis, (b) threefold symmetry axis and (c) twofold symmetry axis.

CHAPTER 13

BIOLOGICAL APPLICATIONS

The golden ratio plays a role in the growth of many biological systems. This is manifested in several distinct ways which can be related to the mathematical properties of the golden ratio which have been discussed in previous chapters. Three particular situations, as follows, will be discussed in this book;

(1) symmetry characteristics,
(2) optimal spacing and
(3) Fibonacci growth spirals.

Biological symmetry

Biological systems exhibit a wide variety of symmetry characteristics. In most cases the symmetry is only approximate as the growth of biological organisms is not perfect and this kind of symmetry is sometimes referred to as *material symmetry*. A number of organisms exhibit either two dimensional fivefold symmetry or three dimensional icosahedral symmetry. These systems have features with dimensions that are related to the golden ratio. In this chapter some of the more common organisms which show fivefold symmetry are described.

Plants

Among the higher plants, the flowering plants are those which display the most obvious fivefold symmetry. The flowering plants, Class *Angiospermae*, are from the phylum *Pteropsida* (which also includes the ferns, Class *Gymnospermae*). The flowering plants are divided into two subclasses; the *Monocotyledonae* and the *Dicotyledonae* (commonly referred to as Monocots and Dicots, respectively). The Monocots have flower petal arrangements with three-, six- or twelvefold symmetry; while the Dicots show four- or fivefold symmetry. The fivefold symmetry which occurs may be either strictly rotational symmetry or both rotational and inversion symmetry, depending on the shape of the petals. Typical examples of the fivefold symmetry which arises in the growth of flowers are illustrated in Fig. 13.1.

(a)

(b)

Fig. 13.1. Flowers with fivefold symmetry; (a) rotational only [*Tabernaemontana corymbosa*] and (b) rotational plus inversion [*Hippobroma longiflora*]. From M. Hargittai (1992), copyright World Scientific Publishing.

Animals

In the animal kingdom, fivefold symmetry is most common among the *Echnodermata*. This phylum includes animals such as starfish, sea urchins and sand dollars and consists almost exclusively of animals which are aquatic. The phylum is divided into four classes, three of which, as described below, show some aspects of fivefold symmetry.

The Class *Crinoidea* includes the crinoids and feather stars, the majority of which are extinct and are only known from the fossil record. Most organisms in this class exhibit fivefold symmetry, or in some cases tenfold symmetry. A typical example is illustrated in Fig. 13.2.

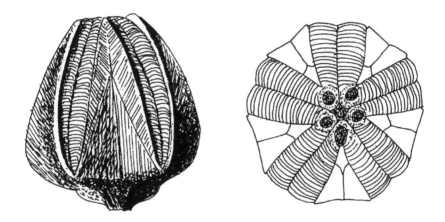

Fig. 13.2. Blastoid [*Pentramites robustus*] from the Class *Crinoidea* showing fivefold symmetry; (a) side view and (b) ventral view.

The class *Stelleroidea* contains two subclasses, the *Asteroidea* or true starfish and the *Ophiuroidea* or brittle stars and their relatives. Fivefold symmetry is abundant in both of these subclasses and a typical example is shown in Fig. 13.3.

Class *Echinoidea* is divided into two subclasses; *Regularia* and *Irregularia*. Many members of the Subclass *Regularia* exhibit fivefold symmetry. These animals include the sea urchins and the sand dollars. Their tests (commonly, but not correctly, referred to as shells) frequently exhibit patterns which show fivefold symmetry as illustrated in Fig. 13.4.

Fig. 13.3. Starfish [*Asterias* sp.] exhibiting fivefold symmetry. From Hartner (1979).

Fig. 13.4. Test of the common sand dollar [*Echinarachnius parma*] showing the fivefold symmetry pattern characteristic of these animals. From Hartner (1979).

Viruses

The viruses contain a proliferation of forms which show symmetry related to the icosahedron. A detailed discussion of the classification of viruses is beyond the scope of this book. However, the symmetry characteristics which are relevant to a discussion of the golden ratio are quite evident in the example shown in Fig. 13.5.

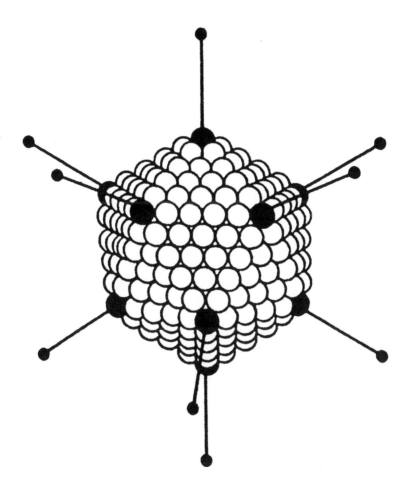

Fig. 13.5. Capsomer of a virus [*Adenovirus*] exhibiting icosahedral symmetry. The diagram shows the protein based structural units known as hexons (open circles) and pentons (solid circles and connecting lines).

Optimal spacing

The optimal spacing of points on the circumference of a circle has been discussed in Chapter 9. There it was shown that angles related to the golden ratio lead to the most uniform spacing of points. This principle is important to the growth of a number of plants. In many cases the arrangement of leaves on the stem of a plant is related to the optimal spacing algorithms discussed in Chapter 9. It is advantageous for leaves to be positioned along the stem in such a manner that will optimize the exposure of each leaf to sun and rain. That is, if the stem is vertical and the leaves are viewed from above, it would be beneficial to avoid the situation where leaves are positioned directly above other leaves. An ideal arrangement is illustrated in Fig. 13.6. Here the growth angle is $2\pi/\tau^2 \approx 137.5077°$. Recall from

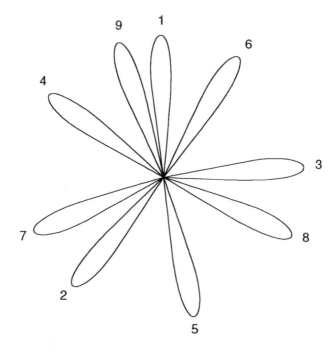

Fig. 13.6. Optimal spacing of leaves on a stem with a growth angle of $2\pi/\tau$ (*radians*). The numbers indicate the order of growth.

Chapter 9 that spacings related to τ^2 are equivalent to those related to τ. In many cases the arrangement of leaves on a plant's stem is related to a rational approximant of the golden ratio; That is growth angles related to $2\pi/\tau_n^2$, where τ_n is the nth rational approximant of τ and is given by a ratio of Fibonacci numbers. Figure 13.7 shows a typical example of this type of growth. If a number of turns, n_1, are taken by an arrangement of n_2 leaves around the stem then the growth angle will be given by $2\pi n_1/n_2$. In the example shown in the figure $n_1 = 3$ and $n_2 = 8$. Both these numbers are Fibonacci numbers, although not successive ones. It is straightforward, however, to show that

$$\lim \frac{F_n}{F_{n-m}} = \tau^m \quad . \tag{13.1}$$

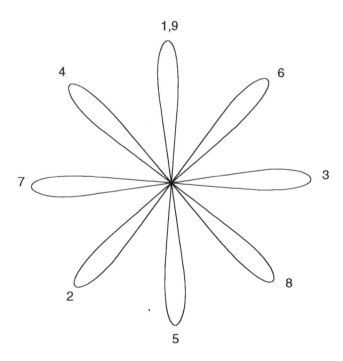

Fig. 13.7. Spacing of leaves on a stem with a growth angle of $2\pi/(8/5)$ (*radians*). Compare with Fig. 13.6.

In the present example 8/3 is the fifth rational approximant of τ^2 and it is clear that in the limit of large n, the situation shown in Fig. 13.6 is achieved. In fact, values of n_1 and n_2 which are Fibonacci numbers are nearly universal among plants which show a regular arrangement of leaves. Typical m in Eq. (13.1) is 2 and the spacing is governed by a growth angle related to a rational approximant of τ^2.

Fibonacci spirals

The concept of Fibonacci spirals in growth patterns is an extension of the ideas described in the above section on optimal spacing. Many organisms show growth patterns which are in the form of spirals. In some cases the spiral nature of the growth is very apparent, as in the case of the shells of most gastropods (snails) and some other mollusks. In other cases the growth of an organism is in the form of discrete components which exhibit a spiral structure. Some examples of this type of structure are the scales on a pine cone or the surface of a pineapple. The golden ratio, or at least ratios of Fibonacci numbers are a common feature of the growth angles of these patterns. Some typical examples of Fibonacci spirals in the growth of different organisms are discussed below.

Pinecones

The pattern on the base of a typical pinecone is illustrated in Fig. 13.8. A spiral arrangement of the seed bearing scales is seen indicating a growth outward from the stem. Both clockwise and counterclockwise spirals can be seen in the figure. It is clear that the clockwise spiral pattern is tighter than the counterclockwise spiral pattern (pine cones of the opposite parity also exist). This indicates that the pattern of scales may be viewed in terms of the outward growth of a series of spirals. The numbers of clockwise and counterclockwise spirals are almost always successive Fibonacci numbers. In the example shown in the figure there are 13 and 8 clockwise and counterclockwise spirals, respectively. This property indicates that the growth angle of the scales in the pinecone is related to a rational approximant of the golden ratio; i.e. 13/8.

Compound Flowers

The heads of many flowers show a complex arrangement of seeds. The sunflower, as illustrated in Fig. 13.9, is a well known example. A spiral pattern consisting of both clockwise and counterclockwise spirals, similar to that displayed by pinecones, is seen in the figure. Typically the numbers of clockwise and counterclockwise spirals are successive Fibonacci numbers and are dependent on the overall size of the flower and the individual seeds. A medium sized sunflower may have a ratio of clockwise to counterclockwise spirals which is 89/55 while a large sunflower would probably have 144/89. In some cases (Hoggatt 1969) a ratio of 123/76 has been observed, indicating a *Lucas sunflower*. These observations indicate a growth angle which is related to τ or at least to a rational approximant of τ. It is interesting to see how important the precise value of the growth angle is in determining the structure of a flower. This problem has been investigated in detail by Rivier *et al.* (1984). Figure 13.10a shows a computer simulated growth pattern for the seeds of a sunflower which has been produced using a growth angle of $2\pi/\tau$. This pattern is generated by determining the angular position of the seeds as $2\pi n/\tau$ (for integer n) and their radial positions on the basis of the previous radius of the flower. The pattern of Fig. 13.10a is similar to that of the real

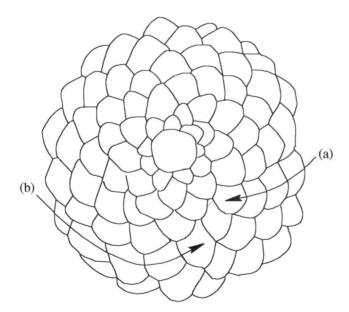

Fig. 13.8. Schematic illustration of the seed bearing scales on the base of a typical pinecone. Line *a* shows one of the 13 clockwise spirals and line *b* shows one of the 8 counter clockwise spirals.

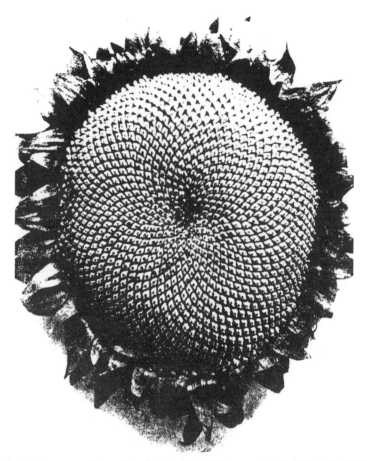

Fig. 13.9. Seed pattern on a sunflower. From Brandmüller (1992), copyright World Scientific Publishing.

sunflower shown in Fig. 13.9. In particular, the arrangement of different numbers of clockwise and counterclockwise spirals is seen. Using the same method of producing a computer generated flower but with a growth angle of $2\pi/(21/13)$, i.e. the eighth rational approximant of τ, the pattern shown in Fig. 13.10b is obtained. This pattern bears little resemblance to the real sunflower. Specifically, the angular position of the seeds takes on a limited number of different values resulting in radial arms in the structure. Since the angular position of the seeds is generated by the expression $2\pi n/(21/13)$ then only 21 different values for the angle can be generated. A reasonable model of a compound flower can be produced only

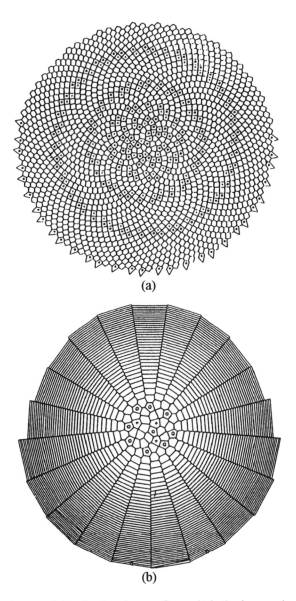

(a)

(b)

Fig. 13.10. Computer generated seed pattern for a sunflower obtained using growth angles of (a) $2\pi/\tau$ (*radians*) (= 222.492°) and (b) $2\pi/(21/13)$ (*radians*) (= 222.857°). Reprinted with permission from Rivier *et al.*, J. Physique **45** (1984) 49.

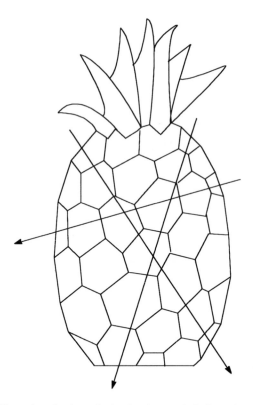

Fig. 13.11. Schematic illustration of a pineapple showing the two clockwise and one counter clockwise sets of spirals. Adapted from Coxeter (1961).

by using a rational approximant of order n, i.e. F_{n+1}/F_n, to give a growth angle $2\pi/(F_{n+1}/F_n)$ where n is sufficiently large that F_{n+1} is at least as large as the number of seeds around the circumference of the flower.

Pineapples

The spiral pattern of the scales on the surface of a pineapple is readily seen if the pineapple is *unrolled* to yield a two dimensional representation as shown in Fig. 13.11. Three sets of spirals are visible and are indicated in the figure. This pattern of scales is readily produced by a computer simulation using a growth angle of $2\pi/\tau_n^2$.

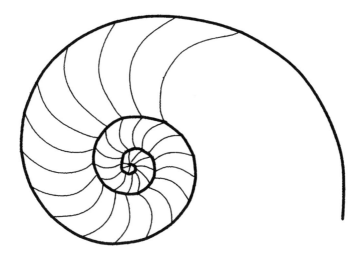

Fig. 13.12. Schematic illustration of the spiral structure of the chambered nautilus [*Nautilus pompilius*]. The internal structure of the growth chambers can be seen.

Mollusks

Probably the most striking example of spiral growth related to the golden ratio in the Phylum *Mollusca* is seen in the chambered nautilus (*Nautilus pompilius*). This animal is illustrated in Fig. 13.12. The nautilus is in the Class *Cephalopoda*, which also contains squid and octopuses. Only a single genus of nautilus is known to exist at present although the fossil record indicates that a proliferation of genera existed during the *Paleozoic* and *Mesozoic* periods. The shell is comprised of a number of chambers and in this way is distinct from the shells of the Subclass *Gastropoda*. As the animal grows it constructs larger and larger chambers in the form of a spiral, sealing off the smaller unused chambers. The shape of the spiral of the chambered nautilus has been considered by Cook (1979). The relative volumes of consecutive chambers is related to the golden ratio. It is generally considered that this quantity is of relevance to the growth of biological organisms because it is the basis of a series which is both arithmetic and geometric in nature.

APPENDIX I

CONSTRUCTION OF THE REGULAR PENTAGON

There are several method of constructing a regular pentagon with a straight edge and a compass. They are all, more or less, equivalent in that they rely upon the construction of two line segments with a length ratio of 1 to τ. The method described below is, therefore, not unique but is representative of this type of construction. The instructions refer to Fig. A1.1.

(1) Construct a square $ABCD$. The edge length of the square, length AB, will be the edge length of the resulting pentagon.

(2) Locate the midpoint of edge AB of the square, labelled E.

(3) Set the compass to draw a circle with a radius equal to the line segment EC and draw an arc, arc 1 in the figure using point E as the center, from point C back down to an extension of the base of the square. This will define point F.

(4) Set the compass to draw a circle with a radius equal to the length AB. Draw an arc, arc 2, with A as the center.

(5) Draw another arc of radius AB, arc 3, with point F as the center. Arcs 2 and 3 will intersect at the point defined as G.

(6) Set the compass to a radius equal to the length AF and draw an arc with G as the center, arc 4. This will intersect arcs 2 and 3 at points H and I, respectively.

(7) Points $AGFIH$ define the vertices of a regular pentagon.

When this construction is performed it is best to make the pentagon as large as possible as this will improve the accuracy in locating the points. It is also important to be careful in setting the compass as small differences in the radii of the arcs will have a fairly substantial effect on the location of points H and I.

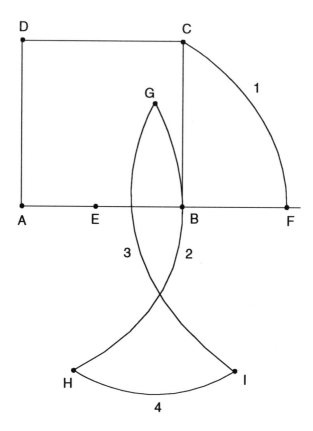

Fig. A1.1: The construction of a regular pentagon.

APPENDIX II

THE FIRST 100 FIBONACCI AND LUCAS NUMBERS

n	F_n	L_n
0	0	2
1	1	1
2	1	3
3	2	4
4	3	7
5	5	11
6	8	18
7	13	29
8	21	47
9	34	76
10	55	123
11	89	199
12	144	322
13	233	521
14	377	843
15	610	1364
16	987	2207
17	1597	3571
18	2584	5778
19	4181	9349
20	6765	15127
21	10946	24476
22	17711	39603
23	28657	64079
24	46368	103682

25	75025	167761
26	121393	271443
27	196418	439204
28	317811	710647
29	514229	1149851
30	832040	1860498
31	1346269	3010349
32	2178309	4870847
33	3524578	7881196
34	5702887	12752043
35	9227465	20633239
36	14930352	33385282
37	24157817	54018521
38	39088169	87403803
39	63245986	141422324
40	102334155	228826127
41	165580141	370248451
42	267914296	599074578
43	433494437	969323029
44	701408733	1568397607
45	134903170	2537720636
46	1836311903	4106118243
47	2971215073	6643838879
48	4807526976	10749957122
49	7778742049	17393796001
50	12586269025	28143753123
51	20365011074	45537549124
52	32951280099	73681302247
53	53316291173	119218851371
54	86267571272	192900153618
55	139583862445	312119004989
56	225851433717	505019158607
57	365435296162	817138163596
58	591286729879	1322157322203
59	956722026041	2139295485799
60	1548008755920	3461452808002
61	2504730781961	5600748293801

62	4052739537881	9062201101803
63	6557470319842	14662949395604
64	10610209857723	23725150497407
65	17167680177565	38388099893011
66	27777890035288	62113250390418
67	44945570212853	100501350283429
68	72723460248141	162614600673847
69	117669030460994	263115950957276
70	190392490709135	425730551631123
71	308061521170129	688846502588399
72	498454011879264	1114577054219522
73	806515533049393	1803423556807921
74	1304969544928657	2918000611027443
75	2111485077978050	4721424167835364
76	3416454622906707	7639424778862807
77	5527939700884757	12360848946698171
78	8944394323791464	20000273725560978
79	14472334024676221	32361122672259149
80	23416728348467685	52361396397820127
81	37889062373143906	84722519070079276
82	61305790721611591	137083915467899403
83	99194853094755497	221806434537978679
84	160500643816367088	358890350005878082
85	259695496911122585	580696784543856761
86	420196140727489673	939587134549734843
87	679891637638612258	1520283919093591604
88	1100087778366101931	2459871053643326447
89	1779979416004714189	3980154972736918051
90	2880067194370816120	6440026026380244498
91	4660046610375530309	10420180999117162549
92	7540113804746346429	16860207025497407047
93	12200160415121876738	27280388024614569596
94	19740274219868223167	44140595050111976643
95	31940434634990099905	71420983074726546239
96	51680708854858323072	115561578124838522882
97	83621143489848422977	186982561199565069121

98	135301852344706746049	302544139324403592003
99	218922995834555169026	489526700523968661124
100	354224848179261915075	792070839848372253127

APPENDIX III

RELATIONSHIPS INVOLVING THE GOLDEN RATIO
AND GENERALIZED FIBONACCI NUMBERS

The relationships given in this appendix provide information concerning the golden ratio as well as Fibonacci, Lucas and generalized Fibonacci numbers. Some of these relationships have been proven in the text while other are provided here without proof. In the following τ is the golden ratio, $\phi = -1/\tau$, α is the Tribonacci number, F_n, L_n and T_n are Fibonacci, Lucas and Tribonacci numbers, respectively, and G_n are generalized Fibonacci numbers. This table is not intended to be comprehensive. Further relations may be found in Vajda (1989).

Fundamental recursion relations

$$F_{n+2} = F_{n+1} + F_n \tag{A3.1}$$

$$L_{n+2} = L_{n+1} + L_n \tag{A3.2}$$

$$T_{n+2} = T_{n+1} + T_n \tag{A3.3}$$

$$G_{n+2} = G_{n+1} + G_n \tag{A3.4}$$

$$F_{-n} = (-1)^{n+1} F_n \tag{A3.5}$$

$$L_{-n} = (-1)^n L_n \tag{A3.6}$$

Relations involving Fibonacci and Lucas numbers

$$F_{2n+1} = F_{n+1}^2 + F_n^2 \tag{A3.7}$$

$$F_{n+2}F_{n-1} = F_{n+1}^2 - F_n^2 \tag{A3.8}$$

$$F_{n+1}F_{n-1} - F_n^2 = (-1)^n \tag{A3.9}$$

$$F_n = F_m F_{n+1-m} + F_{m-1}F_{n-m} \tag{A3.10}$$

$$L_{n+m} + (-1)^m L_{n-m} = L_m L_n \tag{A3.11}$$

$$L_{2n} + 2(-1)^n = L_n^2 \tag{A3.12}$$

$$L_{n-1} + L_{n+1} = 5F_n \tag{A3.13}$$

$$F_{n-1} + F_{n+1} = L_n \tag{A3.14}$$

$$F_{n+2} - F_{n-2} = L_n \tag{A3.15}$$

$$F_n + L_n = 2F_{n+1} \tag{A3.16}$$

$$F_{2n} = F_n L_n \tag{A3.17}$$

$$F_{n+1}L_{n+1} - F_n L_n = F_{2n+1} \tag{A3.18}$$

$$F_{n+m} + (-1)^m F_{n-m} = L_m F_n \tag{A3.19}$$

$$F_{n+m} - (-1)^m F_{n-m} = F_m L_n \tag{A3.20}$$

$$L_m F_n + L_n F_m = 2F_{n+m} \tag{A3.21}$$

$$F_n L_m - L_n F_m = (-1)^m 2F_{n-m} \tag{A3.22}$$

$$L_{n+m} - (-1)^m L_{n-m} = 5F_m F_n \tag{A3.23}$$

$$L_n^2 - 2L_{2n} = -5F_n^2 \tag{A3.24}$$

$$L_{2n} - 2(-1)^n = 5F_n^2 \tag{A3.25}$$

$$5F_n^2 - L_n^2 = 4(-1)^{n+1} \tag{A3.26}$$

$$F_{n+1}L_n = F_{2n+1} - 1 \qquad (n \text{ odd}) \tag{A3.27}$$

$$3F_n + L_n = 2F_{n+2} \tag{A3.28}$$

$$5F_n + 3L_n = 2L_{n+2} \tag{A3.29}$$

$$F_{n+1}L_n = F_{2n+1} + 1 \qquad (n \text{ even}) \tag{A3.30}$$

$$L_n = F_n + 2F_{n-1} \tag{A3.31}$$

$$L_n = F_{n+3} - 2F_n \tag{A3.32}$$

Relationships involving generalized Fibonacci numbers

$$G_{n+m} = F_{m-1}G_n + F_m G_{n+1} \tag{A3.33}$$

$$G_{n-m} = (-1)^m (F_{m+1}G_n - F_m G_{n+1}) \tag{A3.34}$$

$$G_{n+m} + (-1)^m G_{n-m} = L_m G_n \tag{A3.35}$$

$$G_{n+m} - (-1)^m G_{n-m} = F_m (G_{n-1} + G_{n+1}) \tag{A3.36}$$

Relationships involving sums of Fibonacci and related numbers

$$\sum_{i=1}^{n} G_{2i-1} = G_{2n} - G_0 \tag{A3.37}$$

$$\sum_{i=1}^{n} G_i = G_{n+2} - G_2 \tag{A3.38}$$

$$\sum_{i=1}^{n} G_{2i} = G_{2n+1} - G_1 \tag{A3.39}$$

$$\sum_{i=1}^{n} G_{2i} - \sum_{i=1}^{n} G_{2i-1} = G_{2n-1} + G_0 - G_1 \tag{A3.40}$$

$$\sum_{i=1}^{n} \frac{G_{i-1}}{2^i} = \frac{G_0 + G_3}{2} - \frac{G_{n+2}}{2^n} \tag{A3.41}$$

$$\sum_{i=1}^{n} \frac{F_{i-1}}{2^i} = 1 - \frac{F_{n+2}}{2^n} \tag{A3.42}$$

$$\sum_{i=1}^{4n+2} G_i = L_{2n+1} G_{2n+3} \tag{A3.43}$$

$$\sum_{i=1}^{2n} G_i G_{i-1} = G_{2n}^2 - G_0^2 \tag{A3.44}$$

$$\sum_{i=1}^{2n} F_i F_{i-1} = F_{2n}^2 \tag{A3.45}$$

$$\sum_{i=1}^{2n+1} G_i G_{i-1} = G_{2n+1}^2 - G_0^2 - (G_1^2 - G_0 G_2) \tag{A3.46}$$

$$\sum_{i=1}^{2n+1} F_i F_{i-1} = F_{2n+1}^2 - 1 \tag{A3.47}$$

$$\sum_{i=1}^{n} G_{i+2} G_{i-1} = G_{n+1}^2 - G_1^2 \tag{A3.48}$$

$$\sum_{i=1}^{n} G_i^2 = G_n G_{n+1} - G_0 G_1 \qquad \text{(A3.49)}$$

$$\sum_{i=1}^{n} F_i^2 = F_n F_{n+1} \qquad \text{(A3.50)}$$

$$\sum_{i=1}^{\infty} \frac{F_i}{2^i} = 2 \qquad \text{(A3.51)}$$

$$\sum_{i=1}^{\infty} \frac{iF_i}{2^i} = 10 \qquad \text{(A3.52)}$$

$$\sum_{i=1}^{\infty} \frac{iL_i}{2^i} = 22 \qquad \text{(A3.53)}$$

$$\sum_{i=0}^{n} (-1)^i L_{n-2i} = 2F_{n+1} \qquad \text{(A3.54)}$$

$$5\sum_{i=0}^{n} F_i F_{n-i} = (n+1)L_n - 2F_{n+1} = nL_n - F_n \qquad \text{(A3.55)}$$

$$\sum_{i=0}^{n} L_i L_{n-i} = (n+1)L_n + 2F_{n+1} = (n+2)L_n + F_n \qquad \text{(A3.56)}$$

$$\sum_{i=0}^{n} F_i L_{n-i} = (n+1)F_n \qquad \text{(A3.57)}$$

Limiting series ratios

$$\lim_{n \to \infty} \frac{F_{n+1}}{F_n} = \tau \qquad \text{(A3.58)}$$

$$\lim_{n \to \infty} \frac{F_n}{F_{n-m}} = \tau^m \qquad\qquad (A3.59)$$

$$\lim_{n \to \infty} \frac{L_{n+1}}{L_n} = \tau \qquad\qquad (A3.60)$$

$$\lim_{n \to \infty} \frac{G_{n+1}}{G_n} = \tau \qquad\qquad (A3.61)$$

$$\lim_{n \to \infty} \frac{T_{n+1}}{T_n} = \alpha \qquad\qquad (A3.62)$$

Basic relationships involving the golden ratio and related numbers

$$\tau = \frac{\sqrt{5}+1}{2} \qquad\qquad (A3.63)$$

$$\tau^2 - \tau - 1 = 0 \qquad\qquad (A3.64)$$

$$\phi = -\frac{1}{\tau} \qquad\qquad (A3.65)$$

$$\alpha = \frac{1}{3}\left[\left(19+3\sqrt{33}\right)^{1/3} + \left(19-3\sqrt{33}\right)^{1/3} + 1\right] \qquad\qquad (A3.66)$$

$$\alpha^3 - \alpha^2 - \alpha - 1 = 0 \qquad\qquad (A3.67)$$

$$\frac{\tau}{\phi} = -(\tau+1) \qquad\qquad (A3.68)$$

Relationships involving the golden ratio and Fibonacci numbers

$$F_n = \frac{1}{\sqrt{5}}\left(\tau^n - \phi^n\right) \tag{A3.69}$$

$$L_n = \tau^n + \phi^n \tag{A3.70}$$

$$F_n = trunc\left(\frac{\tau^n}{\sqrt{5}} + \frac{1}{2}\right) \qquad n \geq 0 \tag{A3.71}$$

$$L_n = trunc\left(\tau^n + \frac{1}{2}\right) \qquad n > 1 \tag{A3.72}$$

$$F_{n+1} = trunc\left(\tau F_n + \frac{1}{2}\right) \qquad n > 1 \tag{A3.73}$$

$$L_{n+1} = trunc\left(\tau L_n + \frac{1}{2}\right) \qquad n > 3 \tag{A3.74}$$

$$F_{n+1} - \tau F_n = \frac{(-1)^n}{F_{n-1} + \tau F_n} \tag{A3.75}$$

$$t = \prod_{i=1}^{\infty}\left[1 + \frac{(-1)^{i+1}}{F_{i+1}^2}\right] \tag{A3.76}$$

$$G_n = (G_1 - G_0\phi)\frac{\tau^n}{\sqrt{5}} + (G_0\tau - G_1)\frac{\phi^n}{\sqrt{5}} \tag{A3.77}$$

$$\tau = 1 + \sum_{i=2}^{\infty}\frac{(-1)^i}{F_i F_{i-1}} \tag{A3.78}$$

Relationships involving binomial coefficients

$$G_{n+p} = \sum_{i=0}^{n} \binom{n}{i} G_{p-i} \tag{A3.79}$$

$$G_{2n} = \sum_{i=0}^{n} \binom{n}{i} G_i \tag{A3.80}$$

$$G_{p+2n} = \sum_{i=0}^{n} \binom{n}{i} G_{p+i} \tag{A3.81}$$

$$F_{2n+1} = 1 + \sum_{i=0}^{n} \binom{n+1}{i+1} F_i \tag{A3.82}$$

$$G_{p-n} = \sum_{i=0}^{n} (-1)^i \binom{n}{i} G_{n+p-i} \tag{A3.83}$$

$$F_n = \sum_{i=0}^{\infty} \binom{n-i-1}{i} \tag{A3.84}$$

$$\sum_{i=0}^{2n} \binom{2n}{i} F_{2i} = 5^n F_{2n} \tag{A3.85}$$

$$\sum_{i=0}^{2n+1} \binom{2n+1}{i} F_{2i} = 5^n L_{2n+1} \tag{A3.86}$$

$$\sum_{i=0}^{2n} \binom{2n}{i} L_{2i} = 5^n L_{2n} \tag{A3.87}$$

$$\sum_{i=0}^{2n+1} \binom{2n+1}{i} L_{2i} = 5^{n+1} F_{2n+1} \tag{A3.88}$$

$$\sum_{i=0}^{2n} \binom{2n}{i} F_i^2 = 5^{n-1} L_{2n} \tag{A3.89}$$

$$\sum_{i=0}^{2n+1} \binom{2n+1}{i} F_i^2 = 5^n F_{2n+1} \tag{A3.90}$$

$$\sum_{i=0}^{2n} \binom{2n}{i} L_i^2 = 5^n L_{2n} \tag{A3.91}$$

$$\sum_{i=0}^{2n+1} \binom{2n+1}{i} L_i^2 = 5^{n+1} F_{2n+1} \tag{A3.92}$$

REFERENCES

Abramowitz, M. and Stegun, I.A. 1964 *Handbook of Mathematical Functions* (National Bureau of Standards, Washington).

Berg, M. 1966 "Phi, the golden ratio (to 4599 decimal places) and Fibonacci numbers" Fibonacci Quarterly **4** 157-162.

Brandmüller, J. 1992 "Five fold symmetry in mathematics, physics, chemistry, biology and beyond" in I. Hargittai, ed. *Five Fold Symmetry* (World Scientific Publishing, Singapore) pp. 11-32.

Carlitz, L. 1964 "A note on Fibonacci numbers" Fibonacci Quarterly **2** 15-32.

Chorbachi, W.K. and Loeb, A.L. 1992 "An Islamic pentagonal seal" in I. Hargittai, ed. *Five Fold Symmetry* (World Scientific Publishing, Singapore) pp. 283-306.

Cook, T.A. 1979 *The Curves of Life* (Dover Publications, New York).

Coxeter, H.S.M. 1961 *Introduction to Geometry* (Wiley, New York).

Crowe, D.W. 1992 "Albrecht Dürer and the regular pentagon" in I. Hargittai, ed. *Five Fold Symmetry* (World Scientific Publishing, Singapore) pp. 465-488.

Dunlap, R.A. 1990 "Periodicity and aperiodicity in mathematics and crystallography" Sci. Progress (Oxford) **74** 311-346.

Dunlap, R.A. 1992 "Five-fold symmetry in the graphic art of M.C. Escher" in I. Hargittai, ed. *Five Fold Symmetry* (World Scientific Publishing, Singapore) pp. 489-504.

Dunlap, R.A. 1996 "A straightforward method for extracting real roots from polynomial equations of arbitrary order" Mathematical Spectrum **29** 8-9.

Dunlap, R.A. 1997 "Regular polygon rings" Mathematical Spectrum (in press).

Dürer, A. 1977 *The Painters's Manual 1525* W.L. Strauss, ed. (Abaris Books, New York).

Gillings, R.J. 1972 *Mathematics in the Time of the Pharaohs* (Dover Publications, New York).

Gunbaum, B. and Shepherd, G.C. 1987 *Tilings and Patterns* (Freeman, New

York).

Hargittai, I. ed. 1992 *Five Fold Symmetry* (World Scientific Publishing, Singapore).

Hargittai, M. 1992 "Hawaiian flowers with fivefold symmetry" in I. Hargittai, ed. *Five Fold Symmetry* (World Scientific Publishing, Singapore) pp. 529-541.

Hartner, J. 1979 *Animals: 1419 Copyright-Free Illustrations of Mammals, Birds, Fish, Insects, etc.* (Dover Publications, New York).

Hoggatt, V.E., Jr. 1969 *Fibonacci and Lucas Numbers* (Houghton Mifflin Company, Boston).

Holden, A. 1971 *Shapes, Space and Symmetry* (Columbia University Press, New York).

Huntley, H.E. 1990 *The Divine Proportion* (Dover Publications, New York).

Kepler, J. 1619 *Harmonices Mundi Libri Quinque* (Joannes Plancus, Lincii).

Leonardo di Pisa (Fibonacci) 1857 *Scritti di Leonardo Pisano, Bd. I, II Libro Abbaci 1202/1228* (B. Boncomagni, ed. Rome).

March, R.H. 1993 "Polygons of resistors and convergent series" Am. J. Phys. **61** 900-901.

Pacioli, L. 1509 *De Divine Proportione* (Venice).

Penrose, R. 1974 "The role of aesthetics in pure and applied mathematical research" Bull. Inst. Math. Appl. **10** 266-271.

Penrose, R. 1989 "Tilings and quasicrystals, a non-local growth problem?" in M. Jaric, ed. *Introduction to the Mathematics of Quasicrystals* (Academic Press, New York) pp. 53-80.

Pierre, D.A. 1986 *Optimization Theory with Applications* (Dover Publications, New York).

Rivier, N., Occelli, R., Pantaloni, J. and Lissowski, A. 1984 "Structure of Benard convection cells, phyllotaxis and crystallography in cylindrical symmetry" J. Physique **45** 49-63.

Runion, G.E. 1972 *The Golden Section and Related Curiosa* (Scott, Foresman and Company, Glenview).

Schroeder, M.R. 1984 *Number Theory in Science and Communication* (Springer-Verlag, Berlin).

Srinivasan, T.P. 1992 "Fibonacci sequence, golden ratio and a network of resistors" Am. J. Phys. **60** 461-462.

Vajda, S. 1989 *Fibonacci and Lucas Numbers, and the Golden Section* (Ellis Horwood Limited, Chichester).

Verhegen, H.F. 1992 "The icosahedral design of the great pyramid" in I. Hargittai, ed. *Five Fold Symmetry* (World Scientific Publishing, Singapore) pp. 333-360.

Vorobyov, N.N. 1963 *Fibonacci Numbers* (Heath and Company, Boston).

INDEX

A

Additive sequence, 7
Abramowitz, M., 61
Angiospermae, 123
Asteroidea, 125-126

B

Basis, 111
Bees, 37-38
Berg, M., 13
Binary numbers, 71
Binet, J.P.M., 45
Binet formula, 45
Blastoid, 125
Bonaccio, Filius, 35
Bond orientational ordering, 119
Boulder Dam, 4
Brandmüller, J., 1
Buzjani, Abu'l Wafa'al, 108-109

C

Canonical representation, 72ff
Carlitz, L., 55
Cephalopoda, 135
Cheops, 3
Chorbachi, W.K., 108-109
Commensurate projection, 89-90